Ecological Studies

Analysis and Synthesis

Edited by
W.D. Billings, Durham (USA) F. Golley, Athens (USA)
O.L. Lange, Würzburg (FRG) J.S. Olson, Oak Ridge (USA)
H. Remmert, Marburg (FRG)

Volume 69

Ecological Studies

M.L. Harmelin-Vivien   F. Bourlière
Editors

Vertebrates
in Complex Tropical Systems

With 17 Figures

Springer-Verlag
New York Berlin Heidelberg
London Paris Tokyo

Mireille L. Harmelin-Vivien
Centre d'Océanologie de Marseille
Station Marine d'Endoume
F-13007 Marseille, France

François Bourlière
Université René Descartes
Département de Physiologie
F-75007 Paris, France

QL
606
.6
V47
1989

Library of Congress Cataloging-in-Publication Data
Vertebrates in complex tropical systems/edited by M.L. Harmelin
-Vivien and F. Bourlière.
    p. cm. — (Ecological studies; v. 69)
    Bibliography: p.
    Includes index.
    ISBN 0-387-96740-0
    1. Vertebrates—Tropics—Ecology. 2. Vertebrates—Tropics.
I. Harmelin-Vivien, M.L. (Mireille L.) II. Bourlière, François,
1913–   . III. Series.
QL606.6.V47 1989
596'.052623—dc 19   88-16019
                                                                                                    CIP

Typeset by Publishers Service, Bozeman, Montana
Printed and bound by Edwards Brothers, Inc., Ann Arbor, Michigan
Printed in the United States of America

9 8 7 6 5 4 3 2 1

ISBN 0-387-96740-0 Springer-Verlag New York Berlin Heidelberg
ISBN 3-540-96740-0 Springer-Verlag Berlin Heidelberg New York

Introduction

This book addresses the question of what determines species richness in tropical animals by comparing and contrasting the communities of the five major classes of vertebrates in two environments considered to be the most species-rich on Planet Earth—the coral reef and the rainforest. All the contributors were asked to examine how so many species could coexist in such communities and to discuss the ways species assemblages might have evolved over time. Because the coauthors are ecologists, emphasis is quite naturally placed on the first of these two questions, and the factors contributing to the maintenance of α-diversity are discussed at length. However, the question of the very origin of species richness can never be eluded, though it is more an evolutionary problem than an ecological one; it has therefore also been given some attention occasionally. Since we believe that long-term descriptive data and extended field experience are absolutely essential to formulate meaningful questions and propose realistic models, contributors were selected on the basis of their prolonged field practice; all of them actually spent years in the field and/or participated in long-term research programs.

The present volume has its origin in a symposium held on August 15, 1986 at Syracuse, New York, during the Fourth International Congress of Ecology. Although the time at our disposal was far too short to allow for extended discussions, the exchange of views was so lively and stimulating that we immediately agreed to proceed one step further by preparing for publication revised and extended versions of the contributions of the invited speakers. It took a year to materialize our plans. The end result of this process is now in your hands.

The prospective reader will probably wonder why the volume coverage is restricted to vertebrates. There are at least two reasons for this limitation. First, the very same month we were due to meet at Syracuse, another symposium was to be held at Nairobi on insect diversity in the tropics (Mound, 1987), and it was felt that any duplication of efforts should be avoided. The second, and more important, reason for restricting our subject matter to vertebrates was that the ecology of fishes, amphibians, reptiles, birds, and mammals is often known in much greater detail than that of the hundreds of thousands of species of invertebrates. Furthermore, the five major vertebrate classes also differ greatly enough in their type of habitat, physiology, behavior (particularly mobility), and life histories to offer as many opportunities as arthropods for testing the theoretical models of community structure presently in fashion.

Why also limit the discussion to coral reefs and rainforests? First of all, in spite of the obvious differences between aquatic and terrestrial environments, the two ecosystems selected have much in common. Both are reputed to be "aseasonal," or at least submitted to far less drastic seasonal changes of climate than tropical rivers or tropical savannas. This allows the effects of biotic interactions taking place between ecosystem components and community members to be studied more easily. Coral reefs and rainforests share also the distinctive characteristic of being, so to speak, "living landscapes" with very complex architectures essentially made up of modular living organisms (trees and scleractinian corals). Such tridimensional mosaics of habitats provide their host species with a very broad range of living conditions equaled nowhere else. We did not attempt to cover once more the ecology of tropical freshwater fish communities, as this has been most ably done recently by Lowe-McConnell (1987), to whom the reader is referred for comparison with coral reefs.

The book therefore includes seven chapters. The first two deal with the structure of coral fish communities, with a definite emphasis on the Indo-Pacific.

Next comes a well-documented contribution on amphibian and reptile rainforest communities, with particular reference to those of the neotropics—the area where the species diversity of "herptiles" is indeed the highest. This chapter makes a logical transition to the following three, which deal with terrestrial communities of endothermic vertebrates.

The two following chapters deal with the bird communities of tropical rainforests, with examples drawn from three continents. They demonstrate beautifully the specificity of the order, which makes it difficult to envisage general explanations for diversity that apply to all vertebrates. Obviously, the models generated on evidence from bird communities do not necessarily apply to fish, amphibian, and even reptile species assemblages.

The ornithologists' contributions are followed by a short chapter on rainforest mammal communities. Unfortunately, the structure of the mammal species assemblage as a whole has never been studied comprehensively at any tropical site. Only some of its subsets (primary consumers, primates, bats, and some rodents) have sometimes been adequately investigated in this perspective.

Finally, the editors have attempted to "wrap up" together some of their thoughts on the similarities and differences between coral reef and rainforest communities on the one hand, and on how to relate vertebrate diversity to patterns and processes in the lower trophic levels of the ecosystems on the other.

Let's hope that the data presented, and the hypotheses put forward, will stimulate further investigations on this central problem of tropical ecology.

The Editors

References

Lowe-McConnell, R.H. 1987. Ecological Studies in Tropical Fish Communities, Cambridge University Press, Cambridge, UK.

Mound, L.A. (ed.) 1987. Insect diversity in the tropics: A symposium. Biol. J. Linn Soc. 30, 289–356

Contents

Contributors

BOURLIÈRE, FRANÇOIS   Département de Physiologie, Université René Descartes, F–75007 Paris, France

DRISCOLL, PETER V.   Department of Biology and Environmental Science, Queensland Institute of Technology, Brisbane, QLD 4001, Australia

DUELLMAN, WILLIAM E.   Museum of Natural History and Department of Systematics and Ecology, The University of Kansas, Lawrence, Kansas 60045–2454, USA

ERARD, CHRISTIAN   Laboratoire de Zoologie (Mammifères et Oiseaux), Museum National d'Histoire Naturelle, F–75005 Paris, France

HARMELIN-VIVIEN, MIREILLE L.   Station Marine d'Endoume, Centre d'Océanologie de Marseille, F–13007 Marseille, France

KIKKAWA, JIRO   Department of Zoology, University of Queensland, St. Lucia, QLD 4067, Australia

SALE, PETER F.   Department of Zoology, University of New Hampshire, Durham, NH 03824, USA

1. Diversity of the Tropics: Causes of High Diversity in Reef Fish Systems

Peter F. Sale

Introduction

Although it is widely recognized that coral reef environments support exceptionally diverse communities of fishes, the variation in diversity from one region of coral reefs to another is not so well known. At the crudest level we can use records of numbers of species from taxonomic surveys to generate a picture of this variation in diversity. At this regional scale there is at least an order of magnitude variation from locations such as the Philippines with more than 2000 species recorded, to southwestern Florida where as few as 200 species may occur. A profitable task, but one as yet not undertaken for reef fish, would be a detailed determination of this pattern of variation on a global scale. Data available at present are sufficient to state that the center of the Indo-West Pacific region is the region of greatest species richness, with richness declining progressively as one moves away from here. The Caribbean is a lesser center of richness (Sale, 1980). It's important to note that this variation in the richness of fish communities in coral reef habitat blends imperceptibly, at the less diverse end, into that of diverse communities in other types of inshore environment. There is no abrupt change in diversity as one enters coral reef habitat.

Along with substantial variation from region to region, there is variation among habitats within a region in the richness of fish assemblages. This local variability in diversity is somewhat better known than is the global pattern. Those who have documented this variability, however, have been more interested in species composition within habitats rather than in species richness. What is striking is that, on a scale of hundreds or even tens of meters, as one moves from one habitat to another, the composition of fishes present changes very substantially. Data of Goldman and Talbot (1976) are a clear example of this point. They found that fully 49% of species they collected were restricted to only one or another of the five habitat zones into which they divided their reef, while only 7% were ubiquitous. Species richness also varied from habitat to habitat. Some, such as the surge zone habitat, were relatively depauperate and held a relatively low number of "endemic" species (47 species collected, including 3 "endemic" species). Others, such as the leeward reef slopes, were both rich and with high "endemism" (216 and 58 species, respectively). Numerous comparable patterns have been described on other reefs by other authors (Hiatt and Strasburg, 1960; Smith and Tyler, 1972; Clarke, 1977; Ross, 1986).

A theory of diversity should be able to account for three things apparent in the reef fish data. First, how are such rich assemblages of fish species accommodated in coral reef environments? Second, what are the reasons for the differences in diversity (and associated species composition) from one habitat to another at any coral reef? And third, what are the reasons for the considerable variation in diversity from one biogeographic region to another?

While a substantial amount has been learned about the ecology of coral reef fishes in the past 15 years, it will be no surprise to learn that we are still very far from a complete understanding of the processes structuring their communities. It is possible, however, that we are closer to this complete understanding for reef fishes than we are for any other vertebrate group. The major part of what follows will review our understanding of the maintenance of high diversity in local fish assemblages. The chapter concludes with more speculative comments on the patterns of segregation of fish species among habitats, and on the global biogeographic patterns in occurrence of reef fish species.

Traditional Equilibrium Explanations

Traditionally we have sought "explanations" for high diversity in equilibrial models — models in which it is assumed that the current situation in a community is a persistent one, a stable balance unlikely to shift unless pushed by some outside force. This tendency to assume an equilibrium is very widespread in ecology and has only rather recently come under severe questioning (Strong et al., 1984; Diamond and Case, 1986). Furthermore, in assuming an equilibrium, it has been conventional to also assume that the equilibrium is being maintained largely, if not entirely, through the competitive interactions which occur among similar sympatric species. For this reason, I will refer to the general equilibrium model of the community as the competitive paradigm. It can be briefly set out as follows.

In any local community, one expects to find that most species maintain relatively constant abundances because each one is at, or close to, its optimum density for that particular environment, as set by the availability to it of resources. This occurs because each species has sufficient capacity to increase its numbers that the effects on it of predators and diseases (which will reduce its numbers) are quickly compensated for and the population, if temporarily reduced in size, quickly rises to once more press against the limit set by resources. One also expects that the species present will differ from each other to measurable degree in the ways in which they use resources. This is necessary for an equilibrial system because each species must make use of a sufficiently different set of resources in order that competitive exclusion can be avoided. If species are too similar in what they require, competition between them must occur and one or the other will be eliminated from the system. This occurs because one will inevitably be more efficient than the other in using the resources for which they compete.

When we turn specifically to tropical communities, the competitive paradigm is augmented by additional assumptions. The tropics have been viewed as benign climatically, a stable environment, one in which a diverse, biologically accommodated community (cf. Sanders, 1968) should develop. They are more benign, more stable than the temperate zone, and for this reason should support more diverse communities. The differences in resource requirements among species will be very fine, although it is not clear from theory whether they should be highly specialized but distinctive ecologically, or should be more generalized and strongly overlapping in what they do.

Evidence Relevant to This Equilibrium View

What is the evidence in support of this competitive paradigm? It is a widely acknowledged fact that reef fish assemblages do exhibit the expected patterns of resource partitioning, with respect to both food and habitat resources. Beginning with Hiatt and Strasburg's (1960) attempt to document patterns of use of food and space by fish species found at Enewetak atoll, the patterns of dietary or habitat partitioning have been documented many times. Comprehensive dietary surveys have also been done for the fish of southern Japan (Sano et al., 1984a) and the Caribbean (Randall, 1967). Closer analyses of dietary differences have been done for small guilds of species, usually under the assumption that differences in diet are important in explaining the coexistence of the species involved. Thus there are reports for sympatric holocentrids (Gladfelter and Johnson, 1983), chaetodontids (Harmelin-Vivien and Bouchon-Navarro, 1982), and lethrinids (Walker, 1978), for example.

Reports such as that of Goldman and Talbot (1976) discussed above provide similar data on the partitioning of habitat. Frequently, detailed reports on habitat partitioning have been done for small guilds of coexisting species (Waldner and Robertson, 1980; Robertson et al., 1979; Nagelkerken, 1982). Again, the assumption is usually made that the differences in habitat use are somehow of importance in permitting the co-occurrence of the species discussed.

Despite the existence of these detailed studies of resource use, relatively little attention has been directed to whether or not species were actually limited by the availability of their resources. Until recently, virtually nothing was known about the possible limits set by food availability, although Hobson (1974), among others, had argued cogently for the importance of food competition in accounting for the evolution of trophic specializations. By contrast, there was a general appreciation of the probable importance of competition for space which was bolstered by observations of the rapid colonization of novel or denuded structured spaces such as artificial reefs (Talbot et al.,

1978; Molles, 1978; Sale and Dybdahl, 1975, 1978), and by the detailed documentation of strong interspecific defense of territories (Low, 1971; Robertson et al., 1976; Sale, 1977).

The important point to acknowledge here is that the data that were collected on resource partitioning were compatible with the view of the system as competitively regulated and at a stable equilibrium. Each species of fish appeared to be sufficiently distinct in its requirements for resources that permanent coexistence could occur.

Collecting data which are compatible with the predictions of a particular model is not a satisfactory way of testing that model. Furthermore, it is not clear whether any patterns of resource partitioning likely to be derived from empirical research would be incompatible with the competitive paradigm. Roughgarden (1974) in a theoretical paper in which he used some empirical data on diets of reef fish demonstrated that specialized and distinct patterns of resource utilization, such as Randall (1967) had demonstrated in Caribbean serranids, as well as broad, strongly overlapping, but not identical, patterns of resource use, such as represented in the diets of scarids in the Caribbean, were both compatible with the competitive paradigm. This fact, which should have disturbed, did not deter our enthusiasm for collecting such data.

The undeniable fact is that data on the foods eaten, or on the microhabitats used by samples of a number of different species of fish, are inevitably going to differ to some extent among species. Indeed, the more carefully the data are collected, the less likelihood there is that two or more species will be found to be identical in the foods they ate or the habitats they used. Yet any pattern of difference appears to be compatible with the competitive paradigm. One potentially useful approach is to examine whether species differ more than one would expect by chance alone in their use of resources, because this would suggest that divergence in requirements had occurred. This approach has been used occasionally for reef fish (Sale and Williams, 1982), but it has the disadvantages common to all approaches which generate null models from an empirical data set. There are unavoidable implicit assumptions, and it may not be sensitive to pervasive effects of competition in the original data (Hastings, 1987). Under the competitive paradigm, documenting that coexisting species are ecologically different, or even showing that they are more different than one might expect under a competition-free null model, is not as important as is demonstrating that it is these ecological differences which are necessary to permit them to coexist. Attempts to show this for reef fish have been few and, as we will see, most such attempts have led to reconsideration of the paradigm.

Let's examine the evidence that the sizes of reef fish populations are normally set by the availability of their resources. First there is the observation that numbers of fish on a reef don't seem to vary much from year to year. Hiatt and Strasburg (1960, p. 66) came as close as anyone to putting this widely held view on paper when they wrote of

[the coral reef] ecosystem which apparently fluctuates in composition very little, if at all, from year to year, and has over a long period of time acquired a biota successfully adjusted competitively to the relatively constant environment of the tropical west-central Pacific Ocean.

In fact, as I noted previously (Sale, 1980), many reef fish ecologists confidently clung to this view while their own data showed that it was not correct. I had published data on the dispersion of *Dascyllus aruanus*, a small damselfish which occurs in small social groups occupying isolated coral heads of a number of branching species (Sale, 1972). I demonstrated that in most of the shallow sites where I had sampled, the number of fish in a group was strongly correlated with the size of the coral head, and therefore, presumably, with the amount of useful living space it provided. I noted that such a relationship did not hold in a minority of sites, and postulated that in these places living sites were not in short supply. But I completely ignored the broader significance of this observation; on the reef I studied, suitable habitat was apparently in short supply (i.e., limiting to the numbers of fish) only in some local sites. Smith and Tyler (1972, 1975) reported that the assemblage of fish present at particular small patch reefs remained more or less constant in structure over several years, yet their own published data, in addition to indicating some considerable change in species composition from one census to another, also showed some pronounced variation in numbers present of fish of particular species, and of all species combined (Sale, 1980). Ogden and Ehrlich (1977) presented a table documenting numbers of grunts (Haemulidae) present in permanent resting schools on a selection of patch reefs. Yet while they spoke of the constancy of size of these groups, the table showed large variations in size of groups from year to year. The fact is that the "observation" that numbers of fish do not vary greatly from year to year falls apart as soon as detailed data on abundances of species are collected. Apart from the problem of what degree of variation is compatible with "constancy" in the real world, this problem of "seeing" what we expect, rather than what is there, may be of more general importance in ecological research (Sale, 1988).

Clear evidence of temporal variability in community structure is available in a long-term monitoring study in which I have been censusing the assemblages of fish present on 20 small, otherwise undisturbed patch reefs since 1977. I have been able to document pronounced variation in composition and substantial variation in numbers of fish present from census to census (Sale and Douglas, 1984; Sale and Steel, 1986). Over the 10-year period, the total number of fish on any patch reef has varied at least twofold, and in many cases substantially more (Sale and Steel, 1986).

It is always possible that variations in local abundance reflect adjustments of a species (or a community) to track a varying carrying capacity, i.e., fluctuations in numbers do not of themselves mean that each species is not continuously limited by resource availability. On the other hand, when numbers fluctuate by a factor of 2 times or more, when no obvious changes to the resources available have occurred, and particularly when fluctuations occur at different times or proceed in different directions on neighboring sites, tracking of a varying resource base must be considered an unlikely possibility. This is the case for the 20 patch reefs I monitored, as well as for the data of Ogden and Ehrlich (1977), and for some data on pomacentrid fish on small patch reefs collected by Williams (1980). The data on number of fish present on replicate artificial reefs provided by Talbot et al. (1978) illustrate the same phenomenon. The very large variances in their data (see their figure 4) are due to the fact that differing numbers of fish were present on each replicate reef, and numbers on each reef changed from census to census.

A more direct way of testing whether an assemblage of fish is at carrying capacity is to add additional fishes. This is technically difficult at all times and operationally impossible for some age classes or species. Sale (1976) demonstrated reduced survival of added juveniles of a territorial pomacentrid in a habitat in which resident fish main-

tained contiguous territories which carpeted the available space. This study was flawed
by low numbers of fish and only partial replication. Subsequent experimental studies
in which replicate patch reefs were built and populated with varying numbers of
pomacentrids have demonstrated survivorship to be little if at all affected by stocking
rate (over densities spanning the full range of densities seen in natural groups),
although growth rates of juveniles do appear to be negatively related to degree of
crowding (Doherty, 1982, 1983a; Jones, 1987). An alternative experimental approach
is to remove resources from natural assemblages and monitor change in the size of the
populations of fish. Robertson and coworkers did this for territorial space for a
pomacentrid fish (Robertson et al., 1981) and for shelter sites for a labrid (Robertson
and Sheldon, 1979) with little evidence that such space resources directly determined
sizes of populations.

It must be noted that there is increasing evidence that some fish populations suffer
a shortage of food. The effects of crowding on rates of growth in the experiments of
Doherty and Jones seem most easily explained this way, and Jones (1986) recently
documented a direct and dramatic effect of food augmentation on rates of growth and
maturation. Less directly, it has long been noted that many herbivorous reef fish are
interspecifically aggressive principally toward potential competitors for food (Low,
1971; Robertson et al., 1976), and Robertson (1982) even argued that the importance
of fish feces as food for other fish is an indication of the shortage of nutritive food supplies. It appears, however, that the great flexibility of patterns of growth typical of fish
prevent the fluctuations in food supplies which occur having direct effects on survivorship, and therefore on the structure of fish assemblages.

To summarize, the available evidence is not compellingly in favor of the competitive
paradigm as a correct model for reef fish assemblages. While extensive data exist showing a partitioning among reef fishes of food or habitat resources, data showing that this
partitioning is responsible for their coexistence and data confirming that their populations are maintained at constant sizes limited by resources are far less evident.

Further, more detailed analyses of patterns of partitioning will not be useful because
any pattern of resource use other than complete identity is compatible with the competitive paradigm. Complete identity among species becomes a very unlikely result
when data are precisely examined. Nor will further data on trends in sizes of populations help because the "constancy" required by the competitive paradigm tolerates
some variability. Only experimental investigations of the effect of the presence of one
species on another, or of the effects on one or more species of manipulations to
resource supplies, can hope to test this model. Those which have been done so far have
not been supportive but they, and any likely to be done in the future, are necessarily
small-scale manipulations done over short time periods, and can only hope to reject the
model in the particular case.

Peters (1976) claimed that the competitive paradigm was tautological and therefore
not valid science. It certainly is sufficiently flexible to be very difficult to reject, and
will undoubtedly continue to receive some support by those attracted to its self-
consistency (cf. arguments in Anderson et al., 1981, and Sale and Williams, 1982), or
those unwilling to enter a more dynamic world. Nevertheless, for reef fish assemblages
there are alternative explanations available that deserve serious evaluation—

alternatives that are compatible with *all* the evidence available at present. We turn to these next.

Reef Fish Life Histories

It has been long popular to view the fish of a coral reef as comparable to the birds of a tropical forest. Superficially, they are similar in that they move about among the portions of a topographically complex and biotically generated habitat. However, their life histories are very different, and the particular life histories characteristic of reef fishes appear to be a key factor in permitting the high diversity so characteristic of coral reefs. The fish of coral reefs are highly fecund, they produce dispersive pelagic larvae, and they are remarkably sedentary once they complete their larval lives and return to a reef. Let's examine each of these aspects of life history in turn.

If we measure fecundity as the production of fertilized eggs, reef fish, like most fish in other environments, are much more fecund than other vertebrates. The number of eggs per clutch can vary from a hundred or so in some small gobies, pipefish, or wrasses, to hundreds of thousands in typical siganids or acanthurids. At the same time, however, the number of clutches produced each year can range from one or two in species which spawn on a lunar cycle within a short annual season, through clutches produced on a lunar or semilunar cycle over several months, to clutches produced on a daily basis over a major part of the year. A tendency to produce numerous clutches must be considered usual among reef fishes. Thresher (1984) provides details of this variability in fecundity over a wide range of species.

While brotulids and some clinids give birth to live young, and siganids (and possibly some tetraodontids) scatter dermersal eggs which are then abandoned, most reef fish species fall into two main groups (Thresher, 1984). Members of 13 families care for dermersal eggs, and members of 36 families fertilize in mid-water eggs which are themselves pelagic (Thresher, 1984). In both groups, the eggs hatch, usually within 2–3 days, into larvae which are pelagic [among demersal spawners, only the pomacentrid *Acanthochromis polyacanthus* cares for the hatched young (Robertson, 1973)]. The larval phase lasts from about 10 to 100 days, depending on the species. Brothers et al. (1983) provide some typical examples of duration of larval life, derived from reading daily increments in otoliths of newly settled juveniles.

There is abundant evidence that reef fishes export their larvae (or pelagic eggs) quickly from the immediate reef environment, although our knowledge of what happens after that during larval life is very limited. Fish which spawn in mid-water usually rush up from the substratum to release the eggs high above the reef. Many of them spawn on downstream points of a reef, engaging in regular migrations to such points in order to do so. Most of them time their spawning either to sunrise or sunset when predation on the eggs may be least effective, or to the peak of the tide so that currents will take eggs away from the reef (Johannes, 1978; Thresher, 1984).

When the eggs are cared for, either in a dermersal nest (the usual situation) or in the mouth or a pouch belonging to one parent, hatching of the eggs appears timed to coincide with sunset and, often also, with times of strong tidal flow. Where hatchlings have been observed, they have been found to swim strongly upward towards the water's surface and to be positively phototactic (Doherty, 1980, 1983b). Again, these characteristics seem likely to enhance the chance of the hatchlings being transmitted rapidly away from the reef on surface currents. Johannes (1978) argued that avoidance of the intense predation that would occur if eggs and larvae remained close to the reef has been the primary factor selecting for such behavior. It is also probable that larval food supplies are more likely to be encountered away from the natal reef (Doherty et al., 1985).

While relatively little is known about the larval phase of reef fishes, they can be taken in plankton tows, or at surface lights many tens of kilometers from reefs, they are found in the stomachs of mid-water fishes, and they are not found in shallow reef lagoons or otherwise in the immediate vicinity of reefs (i.e., within 10 m of reef surfaces) except when very young or when competent to settle as juveniles. Current patterns in the Great Barrier Reef seem sufficient to transport them hundreds of kilometers during their larval existence unless they have behavioral means for avoiding passive transport (Williams et al., 1984), and similar wide dispersal is to be expected in other reef regions.

Once they settle from the plankton, reef fishes become very sedentary. Sale (1978) showed that reef fish of particular sizes were noticeably more sedentary in habits than terrestrial vertebrates of comparable mass. Among species of 15 cm or less in length, there are numerous examples in which adults spend their lives within 1–2 m^3 of water. Some of these are frankly inquiline, others are territorial on small sites, while still others feed in mid-water on plankton yet remain within a small distance of "home" shelter. Few species of this size would possess home ranges in excess of 200 m^2, and larger species are similarly circumscribed in their movements.

These sedentary habits are all the more remarkable when the longevity of many reef species is taken into account. Until recently there had been little attention given to the survivorship of reef fishes, with the exception of a few of commercial or recreational fishing interest (e.g., Stark and Schroeder, 1970; Thompson and Munro, 1978; Nagelkerken, 1981). Nevertheless, there now exist sufficient data on the subject to indicate that annual species are in a distinct minority and that longevities of a decade or more are not unusual (e.g., Munro and Williams, 1985; Sale et al., 1986).

Consequences of These Life Histories in the Patchy Reef Environment

Reef fishes live out their life histories in a spatially very patchy environment. This patchiness is evident on all spatial scales from the geographic in which reefs are scattered over the tropical ocean, to a scale of meters in which the patchwork of coral colonies of differing morphology is scattered over a noncoral substratum. This patchiness combines with the sedentary behavior of the fishes to permit a high degree of habitat specificity among reef species. Furthermore, each species is distributed as a noncontiguous set of local breeding groups within patches of suitable habitat.

The consequences of these habitat and life history features are that recruitment of larvae to reef-associated populations of juveniles and adults is erratic, and that rates of settlement to local sites are not directly influenced by reproductive success at those locations.

The extended pelagic life makes it unlikely that reproductive activity by members of a local breeding group will play any part in producing subsequent recruits to that same group. Thus local settlement is independent of local reproductive success. The rate of settlement of juveniles to reefs is erratic because only a very small proportion of fertilized eggs survive to settle. We have no direct evidence of the pattern of mortality during larval life, but rates of this mortality can be gauged by a comparison of rates of production of offspring in local sites, and rates of settlement of juveniles to those sites. For example, Doherty (1980, 1982) provided data showing that *Pomacentrus wardi* in the lagoon of One Tree Reef produced 8000 hatchlings per m² of suitable patch reef habitat during a spawning season, but that rate of settlement to that habitat over the same three years was only 0.06 juveniles per m² per year, or 0.0007% of production. This suggests that 99.9993% of larvae die before returning to a reef. *Pomacentrus wardi* is not a notably fecund species, the rates of settlement in the years of Doherty's study were not atypical, and his data were derived from a sufficiently extensive set of patches of suitable habitat that the mean rates of production and settlement can be considered reliable. Similar comparisons could be made for a number of other species for which comparable data exist.

When mortality during the larval phase is well in excess of 99%, as seems the case for reef fishes, and when rates of mortality vary somewhat from time to time, as is to be expected in the real world, it follows that the rate of settlement will show a relatively very much greater degree of variability. This pronounced variation, which is a widespread feature of species with pelagic larvae, will be expressed both in space and in time.

Recently, a number of programs have been undertaken to measure the pattern of settlement or of recruitment of reef species. Note here that settlement to the reef marks the time at which an individual joins, or recruits to, the reef-associated population, but while some studies have sampled intensively enough to record settlement, others have measured recruitment on a monthly or an annual basis. In the latter cases, patterns detected result from the combined effects of larval mortality and dispersal, and of mortality in the days and weeks following settlement but before new recruits were censused. In some species, this juvenile mortality may be quite insignificant (less than 5% per month) while in others, a substantial proportion (>50%) of settlers are lost in the first weeks on the reef (Doherty and Sale, 1986; Sale and Ferrell, in press).

Studies of settlement and recruitment have documented the expected variability for a number of species, and on a number of spatial and temporal scales (Victor, 1982, 1986; Sale et al., 1984a; Sale, 1985; Schroeder, 1985). At One Tree Reef, various workers documented for various species several-fold variation in the density of new recruits to local sites 10–10³ m apart (Doherty, 1980; Williams, 1980; Williams and Sale, 1981; Eckert, 1985; Sale et al., 1986; Fowler, 1987). In the same region, order-of-magnitude variation in density of annual recruitment has been documented on a scale of tens of kilometers among neighboring reefs (Eckert, 1984; Sale et al., 1984a). At both scales, patterns for different species often differ substantially.

These same studies have also documented substantial temporal variation both on a daily or weekly scale, and between years. Within a season, local settlement is often markedly episodic (Sale, 1985; Schroeder, 1985; Victor, 1986; Walsh, 1987). Also, the variation among successive seasons often interacts with the variation among sites, so that it has not been yet possible to demonstrate that certain locations are consistently better than others for settlement or recruitment of particular species (Williams and Sale, 1981; Eckert, 1984; Sale et al., 1984a). Patterns of settlement, and perhaps of postsettlement survival, are complicated at the very local scale by microhabitat preferences (Sale et al., 1984b) and by responses shown toward conspecifics by at least some species (Sweatman, 1983, 1985).

Importance of the Temporal Variability of the Reef Environment

While ecologists have tended to think of coral reefs (and other tropical environments) as temporally constant places in which to live, it is becoming apparent that a variety of changes, operating on a range of time scales, cause coral reefs to be temporally variable places to a degree which is important for resident fishes. (It remains possible that they are less changeable environments than are many temperate ones, but even this is not certain when one remembers to factor in the susceptibility of resident species to the changes which occur.) A range of destructive processes occur on coral reefs, including the actions of boring and other bioeroding organisms, "normal" wave action and storm surges, *el nino* phenomena, and outbreaks of disease organisms (Connell, 1978; Woodley et al., 1981; Glynn, 1983; Lessios et al., 1984). Aggregations of the predatory starfish *Acanthaster planci* have warranted particular attention (Potts, 1982; Done, 1985). In addition, the growth of coral is a constructing process which can result in recovery of general appearance of an area of reef within a decade of severe destruction by a starfish outbreak. On longer time scales, changes in sea level and isostatic movements have ensured that all reefs have undergone profound change over geologic time (Davies and Montaggioni, 1985; Digerfeldt and Hendry, 1987). For example, the southern Great Barrier Reef has been a dissected limestone plateau subject to subaerial erosion for most of the past 100,000 years. Water returned about 9000 years ago, but the reefs existed as upward-growing platforms and pinnacles until about 4000 years ago, when upward growth was halted as the reefs reached the no-longer-rising sea surface. Lateral growth to leeward has produced the range of distinctive habitats characteristic of, for example, One Tree Reef, only since that time (Marshall and Davies, 1982; Davies and Montaggioni, 1985).

That reefs are changeable environments on these various temporal scales makes even less tenable the concept of a fish community organized by competitive processes around a relatively constant equilibrium structure. Along with direct effects on fish populations of the various disturbances which occur (Bohnsack, 1983; Lassig, 1983; Sano et al., 1984b; Williams, 1986), there are indirect effects due to the modification of the reef environment. If species differ in their capabilities to occupy the various habitats provided by reefs, and if the habitats are subject to change, then the environment must favor different species to different extents at different times. Competitive outcomes must correspondingly vary. On the other hand, this variability of the environment simply adds to the possibilities for long-term persistence of a diverse array of species as a nonequilibrial system. We will turn next to a brief examination of nonequilibrial mechanisms.

Nonequilibrial Mechanisms for Maintaining High Diversity

Over the past 15 years, empirical and theoretical ecologists have generated a number of hypotheses for the successful long-term coexistence of similar species in a non-equilibrial system. In being nonequilibrial, these hypotheses do not claim to predict a permanent pattern of relative abundance of the coexisting species, nor that the coexistence will be permanent either locally or globally. But each of them provides mechanisms which buffer species from the risk of global extinction when their numbers fall to low levels. The hypotheses which have been developed are not mutually exclusive, and many of them apply easily to the kind of species represented by reef fishes and the kind of habitat provided by coral reefs.

Caswell (1978) developed a simulation model for competition and predation in a patchy environment. He showed that under a wide range of conditions, a predator that fed indiscriminately upon either of two unequal competitors could substantially extend the time to global extinction of the inferior competitor. The important characteristics of his model for permitting this were the extents to which each species was capable of dispersing from one patch to another.

Huffaker (1958) had much earlier explored the dynamics of a predator–prey system in a patchy environment by a direct empirical experiment with insect pests and predators. More recently, Root and Kareiva (1984) did a similar thing with elegant field experiments, but Caswell's study was more complicated in that it looked at the effect of a predator on the outcome of competition among its prey.

Den Boer (1968, 1981) pointed to the widespread phenomenon of spreading of risk, whereby organisms occupying a patchy and changeable environment optimize their chance of reproductive success by dispersing their reproductive effort in time and space, thereby increasing the chance that some offspring will be successful. A variety of reproductive behaviors serve to achieve this risk spreading. Among them are itero-pariety, scattering of eggs when laying, possession of mechanisms for dispersing juvenile or egg stages, and high fecundity. Species which engage in risk-spreading behavior are preadapted for persisting as fugitive species in the presence of superior competitors, or in the presence of competent predators.

The fugitive strategy was first proposed by Hutchinson (1957) as a (rare) phenomenon which permitted the apparent violation of the competitive exclusion principle. A fugitive species occupies a changeable habitat in which new patches, devoid of its competitors, are continually opened up for colonization. The fugitive species is invariably eliminated locally through interactions with its superior competitors but is particularly adept at colonizing such open patches. It is able to persist globally by virtue of the fact that it usually gets to new patches, matures, and reproduces successfully before being excluded from those patches by its competitors. This strategy can also operate when a favored prey species faces intense but patchy predation.

The fugitive strategy now appears to be far more common than Hutchinson supposed, and complete guilds of fugitives have been described (Platt, 1975; Platt and Weis, 1985). Behavior useful for ensuring spreading of risk is the type required for a species to persist as a fugitive. For a fugitive, local persistence is always short term, and competitive or predatory interactions always eliminate it. However, long-term persistence globally is possible because new opportunities in the absence of competitors or predators are continually being created.

Connell (1978) maintained that the rate and severity of disturbances acting on a community can have a major impact on the diversity of that community. In Connell's hypothesis there is an intermediate level of disturbance at which the richest diversity will develop. Systems disturbed to a lesser extent will be less diverse because superior competitors will have eliminated many inferior species. Systems disturbed to a greater degree will be less diverse because only a small number of resistant and quickly recolonizing species will be able to live and reproduce there. Disturbances may be caused by physical or weather events, or they may be due to the action of predators that have a spatially defined impact on their prey communities. Paine and Levin (1981) and Sousa (1981, 1984) most clearly demonstrated the importance of local, spatially defined disturbance regimes in determining the structure of communities.

All these hypotheses are mechanisms for avoiding the inevitable effects of persistent competitive interactions. Sale (1977) and, subsequently, Chesson and Warner (1981; Chesson, 1985, Warner and Chesson, 1985) proposed mechanisms which permit coexistence despite continuing competition. Their mechanisms apply where species compete for living sites or other resources that become available at times or places that cannot reliably be predicted by breeding individuals, and where an individual of either species is capable of defending its site, or other resource, once it has successfully obtained it. In such cases, competition will take on the aspects of a lottery and competitive exclusion will be very unlikely to occur. In such circumstances, species must adopt a strategy of producing numerous dispersive larvae, abundant and variably dormant seeds, or other lottery tickets. Priority of access will determine which species occupies which particular site or other resource, and therefore the species composition at any particular time and place. Chesson (1985) in particular has emphasized the importance of overlapping generations to the success of such a system. These function to store the success of previous bouts of reproduction against an uncertain future. With nonoverlapping generations, lottery competitive systems rapidly lead to loss of species through competitive exclusion, but with the storage effect operating very long-term (essentially permanent) persistence is possible, despite continuous interspecific competition.

It should be clear from what has been said earlier about the life cycles of reef fish and the nature of the habitat in which they live that their communities provide a good example of the kind of system in which all of the mechanisms discussed above might operate. It should also be clear that the various mechanisms outlined are overlapping, and certainly compatible with each other.

Species of reef fish are relatively sedentary and habitat specialists as juveniles and adults, but they have numerous dispersive larvae. Thus our first conclusion must be that while competitive interactions *may* take place within a patch of habitat and *may* lead to the early death of individuals of disadvantaged species, those interactions will not reduce the chance of subsequent settlement to that patch of habitat by new juveniles of that species. Second, there is accumulating evidence of substantial storage of successful recruitment because of the combination of frequent reproduction and relative longevity of many species. And there is evidence of very variable recruitment, so that priority of access to sites is largely a matter determined by factors quite outside the power of the fish to control or even to predict when timing their reproduction. These two factors make possible the operation of lotteries whenever competition occurs. Indeed, I now find it very difficult to visualize a circumstance in which competition among species of coral reef fish could ever lead to exclusion.

The realization that coral reefs are subject to a range of disturbances, from the predator which removes a single site-attached fish to the major physical or climatic event operating on larger temporal or spatial scales, provides for the possibility of disturbance-mediated coexistence. The accumulating evidence that reef fish populations are often not at any limit set by the availability to them of food or space resources, and that population sizes fluctuate markedly through time, is further argument in favor of the operation of nonequilibrial patch-dynamical mechanisms for maintaining the structure of assemblages to which they belong. Under these circumstances, competition between species may be a much more intermittent phenomenon than once thought.

The conclusion has to be that long-term coexistence of numerous similar species of reef fishes is both to be expected and explainable by one or more of the array of mechanisms I have outlined. The high diversity of coral reef fish assemblages is not a problem requiring novel ecological understanding. Neither are present day ecological interactions among fish species responsible for determining local diversity. The patterns of resource partitioning to which we have given so much attention are neither important in permitting coexistence nor likely to be similar from time to time or place to place. Far from being stable, competitively organized systems, reef fish assemblages appear to be groups of species that are independently and successfully gambling to persist in a patchy, changeable world.

Other Patterns of Diversity

Let us turn now from considering the structure of local assemblages and examine the other two features of the distribution of reef fish I identified at the start. Can we understand the patterns of habitat differentiation so evident among reef fishes at the local scale? And can we account for the particular diversity in a region and for the differences in diversity among coral reef regions around the globe? The short answer to both questions is, "No, not yet." A longer, but decidedly speculative answer follows. It is a predominantly evolutionary rather than ecological one.

The nonequilibrial mechanisms discussed above can account for the long-term persistence of species which share a local site, but they cannot account for why different species and different numbers of species share different types of site, nor for why some geographic regions contain many more species than do others. At present our data are very limited, and the careful comparisons have yet to be done. But life cycles, fecundity, and longevity do not appear to differ among geographic regions. Nor do disturbance regimes.

Let's consider the segregation of species among local habitats first. As noted earlier, this segregation is quite striking in all coral reef systems. Few species are such complete habitat generalists that they can be expected to be found everywhere on a reef. What is most common is that as one moves from one reef habitat to another, there is, to a greater or lesser extent, a replacement of one species by another from the same guild. They are sometimes taxonomically quite closely related as well (Clarke, 1977; Sale,

1977, 1980; Waldner and Robertson, 1980). The usual explanation for this is that inter-specific competitive interactions have led to the pattern of habitat partitioning observed; however, as I have discussed, evidence that this is the case is lacking. An alternative explanation may have to do with the evolution of optimal strategies for life.

Reef fish are forced, by the nature of their environment, to disperse numerous off-spring widely in space and time. Each offspring has only a minute chance of finding its way back to a reef, but must be capable of surviving in the reef habitat it finds at settle-ment. There are no second chances except in those cases where the juvenile habitat differs from that used by the adult, such as occurs among some species in the Caribbean which settle initially into shallow sea grass beds. Even here, the juvenile moves a rela-tively short distance from its juvenile to an adjacent adult habitat (Shulman, 1985). Survival in either case requires being able to cope with the physical environment present and with the interactions with other species. Under such circumstances we might expect species to narrow the range of environments that they will colonize in order to be more closely adapted to surviving once settlement has occurred.

Two mechanisms for this evolution of specialization can be envisaged. Both rely on differential mortality of fish across habitats to select for phenotypes well adapted to the particular conditions at each site. Mutations favoring selection of particular habitats at the time of settlement will also be selected for when linked with alleles of adaptive value in that habitat.

The first mechanism assumes conventional allopatric speciation (Bush, 1975), notes the dramatic structural changes that have occurred in all coral reef regions over evolu-tionary time scales, and assumes that different habitats have been particularly exten-sive in different widely separated locations in the past. At each such location (say, a region of extensive shallow lagoons and back reefs, or a region of narrow fringing reefs lacking lagoon development) initially generalized members of a species will be selected for adaptations favored in that habitat and for behavior aiding selection of that habitat at the time of settlement. At some later time, when sea level fluctuations or other changes mix fish with members of the same parental species that have been selected in different habitats in other regions, the presently observed pattern of habitat partition-ing will appear.

In view of the extensive dispersal of larval reef fishes, conventional allopatric mechanisms of speciation seem to me to require unrealistically frequent and major changes to geographic distributions of species populations and of reef habitats. Alterna-tive mechanisms should not be dismissed out of hand. My second mechanism accepts the possibility of parapatric speciation; the evolution of a generalized species into two or more daughter species occupying nearby, different habitats. Parapatric speciation is not widely accepted among evolutionary biologists, although there are some advocates (Bush, 1975; Endler, 1977).

Parapatric speciation of an initially generalized species might occur because selec-tion within each reef habitat will favor certain phenotypes, and because breeding is within habitats, thus favoring assortative mating. Depending on where and when muta-tions happen to occur, the initially generalized species will evolve into one or more populations each with adaptations favoring survival in, and adaptations favoring selec-tion of, a particular subset of the range of habitats offered by the reef. Each daughter population will evolve through this process of increasing habitat specialization without the necessity of any interactions with other populations or species at all. If parapatric speciation does occur, guilds of taxonomically closely related species that replace one another across habitats should be expected. Many examples of such guilds exist among the pomacentrids, the labrids, the acanthurids, and other speciose families.

From these evolutionary considerations, we can make some predictions about the nature of reef fish communities. The evolution of habitat specialization, whether allopatric or parapatric, will be counteracted by the requirement that offspring remain sufficiently generalized in their habitat requirements that their chance of finding suitable sites when settling from the plankton is not made too small. In short, species will not be completely generalized because of a natural tendency to fix new alleles favoring survival in, and favoring selection of, a subset of the wide range of different environments offered by most reefs. But they will not become narrowly specialized because the narrower their requirements at the time of settlement, the smaller the chance of finding a suitable place to live. Measuring the degree of specialization is ecologically a difficult thing to do; however, I argued some years ago that reef fish are no more specialized than are similar kinds of fishes from temperate habitats (Sale, 1977). The literature contains some examples of highly specialized reef species (e.g., Greenfield and Greenfield, 1982), but many that are not, and the habitat preferences of settling fish appear to overlap considerably (Sale et al., 1984b).

Again, regardless of the evolutionary mechanism, the development of adaptations to habitats will depend on chance production of new alleles. Thus, the development by one species of specializations for life in a particular habitat will occur quite independently of the development by other unrelated species of specializations for the same habitat. Numerous, ecologically similar species may occupy the same habitat, and species will occupy more similar habitats than they would if interspecific competition were responsible for the habitat specialization that develops. I previously presented examples of both phenomena (Sale, 1977; Sale and Dybdahl, 1975; Sale and Williams, 1982).

Finally, it seems reasonable to suppose that this process of increasing specialization toward subsets of habitats available should have progressed further in those geographic regions where reefs have been present for longest and in those regions where the physical environments offered by the various habitats are most strikingly different. If the evolution has been primarily allopatric, it should have proceeded further in those regions offering the greatest scope for this. Fish inhabiting the Caribbean, with its small extent, its modest tides, and its generally peaceful seas, may be expected to show a lesser degree of habitat partitioning than is the case in some parts of the Pacific where tides are more extensive and seas are generally rougher. Data addressing this prediction seem not yet to have been collected. Their analysis will be made difficult by the imprecise meaning of the term "specialized."

This leads us to the geographic patterns. It is probable that modern ecological interactions play no real role in determining these. Latitudinal clines in conditions have little if anything to do with the diversity of fish communities on coral reefs. Instead, there is a strong cline of increasing diversity toward the center of the Indo-West Pacific (Sale, 1980). The cline is pronounced and is shared by other marine biota.

The high diversity of reef fish communities is ultimately due to the evolution of large numbers of species, and the regional differences in diversity are a consequence of differing rates of speciation, or of extinction, or of a longer time for these processes to have gone on. Deeper understanding is needed of why groups speciate and why there are the numbers of species that there are. What factors exist which can lead to assortative mating reducing gene flow, especially parapatrically? Is speciation necessarily allopatric and, if so, what mechanisms have operated to achieve isolation in marine species with dispersive larvae? What are the geologic histories of different geographic regions, and do they suggest different patterns of speciation or extinction? Coral reef systems, because they are so easily manipulated, have proved to be a useful system for experimentally exploring ecological questions. Perhaps they will also be a useful system for these evolutionary explorations.

Acknowledgments. The author benefited from the comments of M.J. Caley and R. Shine in preparing this manuscript. Original work by the author and coworkers referred to here was supported by grants from the Australian Research Grants Committee, the Marine Science and Technologies Grants Scheme, the Great Barrier Reef Marine Park Authority, and the University of Sydney. This is a publication from the University of Sydney's One Tree Island Field Station.

References

Anderson, G.R.V., Ehrlich, A.H., Ehrlich, P.R., Roughgarden, J.D., Russell, B.C., and Talbot, F.H. 1981. The community structure of coral reef fishes. Am. Nat. 117, 476–495

Bohnsack, J.A. 1983. Resiliency of reef fish communities in the Florida Keys following a January 1977 hypothermal fish kill. Env. Biol. Fish. 9, 41–54

Brothers, E.B., Williams, D.McB., and Sale, P.F. 1983. Length of larval life in twelve families of fishes at "One Tree Lagoon," Great Barrier Reef, Australia. Mar. Biol. 76, 319–324

Bush, G. 1975. Modes of animal speciation. Annu. Rev. Ecol. Syst. 6, 339–364

Caswell, H. 1978. Predator-mediated coexistence: A nonequilibrium model. Am. Nat. 112, 127–154

Chesson, P.L. 1985. Coexistence of competitors in spatially and temporally varying environments: A look at the combined effects of different sorts of variability. Theor. Pop. Biol. 28, 263–287

Chesson, P.L., and Warner, R.R. 1981. Environmental variability promotes coexistence in lottery competitive systems. Am. Nat. 117, 923–943

Clarke, R.D. 1977. Habitat distribution and species diversity of chaetodontid and pomacentrid fishes near Bimini, Bahamas. Mar. Biol. 40, 277–289

Connell, J.H. 1978. Diversity in tropical rain forests and coral reefs. Science 199, 1302–1310

Davies, P.J., and Montaggioni, L. 1985. Reef growth and sea level change: the environmental signature. Fifth Int. Coral Reef Congr. Proc. 3, 477–511

Den Boer, P.J. 1968. Spreading of risk and stabilization of animal numbers. Acta Biotheor. 18, 165–194

Den Boer, P.J. 1981. On the survival of populations in a heterogeneous and variable environment. Oecologia 50, 39–53

Diamond, J.M., and Case, T.J. (eds.) 1986. Community Ecology. Harper and Row, New York

Digerfeldt, G., and Hendry, M.D. 1987. An 8000 year Holocene sea-level record from Jamaica: Implications for interpretation of Caribbean reef and coastal history. Coral Reefs 5, 165–169

Doherty, P.J. 1980. Biological and physical constraints on the population of two sympatric territorial damselfishes on the southern Great Barrier Reef. Unpub. Ph.D. dissertation, University of Sydney

Doherty, P.J. 1982. Some effects of density on the juveniles of two species of tropical, territorial damselfish. J. Exp. Mar. Biol. Ecol. 65, 249–261

Doherty, P.J. 1983a. Tropical territorial damselfishes: Is density limited by aggression or recruitment? Ecology 64, 176–190

Doherty, P.J. 1983b. Diel, lunar and seasonal rhythms in the reproduction of two tropical damselfishes: *Pomacentrus flavicauda* and *P. wardi*. Mar. Biol. 75, 215–224

Doherty, P.J., and Sale, P.F. 1986. Predation on juvenile coral reef fishes: An exclusion experiment. Coral Reefs 4, 225–234

Doherty, P.J., Williams, D.McB., and Sale, P.F. 1985. The adaptive significance of larval dispersal in coral reef fishes. Env. Biol. Fish. 12, 81–90

Done, T.J. 1985. Effects of two *Acanthaster* outbreaks on coral community structure: The meaning of devastation. Fifth Int. Coral Reef Congr. Proc. 5, 315–320

Eckert, G.J.1984. Annual and spatial variation in recruitment of labroid fishes among seven reefs in the Capricorn/Bunker Group, Great Barrier Reef. Mar. Biol. 78, 123–127

Eckert, G.J. 1985. Population studies on labrid fishes on the southern Great Barrier Reef. Unpub. Ph.D. dissertation, University of Sydney

Endler, J.A. 1977. Geographic Variation, Speciation, and Clines. Princeton Univ. Press, Princeton, NJ

Fowler, A.J. 1987. The development of sampling strategies for population studies of coral reef fishes: A case study. Coral Reefs 6, 49–58

Gladfelter, W.B., and Johnson, W.S. 1983. Feeding niche separation in a guild of tropical reef fishes (Holocentridae). Ecology 64, 552–563

Glynn, P.W. 1983. Extensive "bleaching" and death of reef corals on the Pacific coast of Panama. Env. Conserv. 10, 149–154

Goldman, B., and Talbot, F.H. 1976. Aspects of the ecology of coral reef fishes. In: Jones, O A., and Endean, R. (eds.), Biology and Geology of Coral Reefs, Vol. 3, Academic Press, New York, pp. 125–154

Greenfield, D., and Greenfield, T.A. 1982. Habitat and resource partitioning between two species of *Acanthemblemaria* (Pisces: Chaenopsidae), with comments on the chaos hypothesis. Smithson. Contrib. Mar. Sci. 12, 499–507

Harmelin-Vivien, M.L., and Bouchon-Navarro, Y. 1982. Trophic relationships among chaetodontid fishes in the Gulf of Aqaba. Fourth Int. Coral Reef Symp., Manila 1981, Proc. 2, 538–544

Hastings, A.1987. Can competition be detected using species co-occurrence data? Ecology 68, 117–123

Hiatt, R.W., and Strasburg, D.W. 1960. Ecological relationships of the fish fauna on coral reefs of the Marshall Islands. Ecol. Monogr. 30, 65–127

Hobson, E.S. 1974. Feeding relationships of teleostean fishes on coral reefs in Kona, Hawaii. Fishery Bull. 72, 915–1031

Huffaker, C.B. 1958. Experimental studies on predation: dispersion factors and predator-prey oscillations. Hilgardia 27, 343–383

Hutchinson, G.E. 1957. Concluding remarks. Cold Spr. Harbor Symp. Quant. Biol. 22, 415–427

Johannes, R.E. 1978. Reproductive strategies of coastal marine fishes in the tropics. Env. Biol. Fish. 3, 65–84

Jones, G.P. 1986. Food availability affects growth in a coral reef fish. Oecologia 70, 136–139

Jones, G.P. 1987. Competitive interactions among adults and juveniles in a coral reef fish. Ecology 68, 1534–1547

Lassig, B.R. 1983. The effect of a cyclonic storm on coral reef fish assemblages. Env. Biol. Fish. 9, 55–64

Lessios, H.A., Cubit, J.D., Robertson, D.R., Shulman, M.J., Parker, M.R., Garrity, S.D., and Levings, S.C. 1984. Mass mortality of *Diadema antillarum* on the Caribbean coast of Panama. Coral Reefs 3, 173–182

Low, R.M. 1971. Interspecific territoriality in a pomacentrid reef fish, *Pomacentrus flavicauda* Whitley. Ecology 52, 648–654

Marshall, J.F., and Davies, P.J. 1982. Internal structure and holocene evolution of One Tree Reef, southern Great Barrier Reef. Coral Reefs 1, 21–28

Molles, M.C. 1978. Fish species diversity on model and natural reef patches: Experimental insular biogeography. Ecol. Monogr. 48, 289–305

Munro, J.L., and Williams, D.McB. 1985. Assessment and management of coral reef fisheries: Biological, Environmental and socio-economic aspects. Fifth Int. Coral Reef Congr., Proc. 4, 545–578

Nagelkerken, W.P. 1981. Distribution and ecology of the groupers (Serranidae) and snappers (Lutjanidae) of the Netherlands Antilles. Publ. Found. Sci. Res. Surinam and Netherl. Antilles 107, 1–71

Nagelkerken, W.P. 1982. Distribution of the groupers and snappers of the Netherlands Antilles. Fourth Int. Coral Reef Symp., Manila, 1981, Proc. 2, 480–484

Ogden, J.C., and Ehrlich, P.R. 1977. The behavior of heterotypic resting schools of juvenile grunts (Pomadasyidae). Mar. Biol. 42, 273–280

Paine, R.T., and Levin, S.A. 1981. Intertidal landscapes: Disturbance and the dynamics of pattern. Ecol. Monogr. 51, 145–178

Peters, R.H. 1976. Tautology in evolution and ecology. Am. Nat. 110, 1–12

Platt, W.S. 1975. The colonization and formation of equilibrium plant species associations on badger disturbances in a tall-grass prairie. Ecol. Monogr. 45, 285–305

Platt, W.S., and Weis, I.M. 1985. An experimental study of competition among fugitive prairie plants. Ecology 66, 708–720

Potts, D.C. 1982. Crown-of-thorns starfish: Man-induced pest or natural phenomenon? In: Kitching, R., and Jones, R.E. (eds.), The Ecology of Pests, CSIRO, Melbourne, pp. 54–86

Randall, J.E. 1967. Food habits of reef fishes of the West Indies. Stud. Trop. Oceanogr. 5, 665–847

Robertson, D.R. 1973. Field observations on the reproductive behaviour of a pomacentrid fish, *Acanthochromis polyacanthus*. Z. Tierpsychol. 32, 319–324

Robertson, D.R. 1982. Fish feces as fish food on a pacific coral reef. Mar. Ecol. Prog. Ser. 7, 253–265

Robertson, D.R., Hoffman, S.G., and Sheldon, J.M. 1981. Availability of space for the territorial Caribbean damselfish *Eupomacentrus planifrons*. Ecology 62, 1162–1169

Robertson, D.R., Polunin, N.V.C., and Leighton, K. 1979. The behavioral ecology of three Indian Ocean surgeonfishes (*Acanthurus lineatus*, *A. leucosternon* and *Zebrasoma scopas*): Their feeding strategies, and social and mating systems. Env. Biol. Fish. 4, 125–170

Robertson, D.R., and Sheldon, J.M. 1979. Competitive interactions and the availability of sleeping sites for a diurnal coral reef fish. J. Exp. Mar. Biol. Ecol. 40, 285–298

Robertson, D.R., Sweatman, H.P.A., Fletcher, E.A., and Cleland, M.G. 1976. Schooling as a mechanism for circumventing the territoriality of competitors. Ecology 57, 1208–1222

Root, R.B., and Kareiva, P.M. 1984. The search for resources by cabbage butterflies (*Pieris rapae*): Ecological consequences and adaptive significance of Markovian movements in a patchy environment. Ecology 65, 147–165

Ross, S.T. 1986. Resource partitioning in fish assemblages: A review of field studies. Copeia 1986, 352–388

Roughgarden, J.D. 1974. Species packing and the competition function with illustrations from coral reef fish. Theor. Pop. Biol. 5, 163–186

Sale, P.F. 1972. Influence of corals in the dispersion of the pomacentrid fish, *Dascyllus aruanus*. Ecology 53, 741–744

Sale, P.F. 1976. The effect of territorial adult pomacentrid fishes on the recruitment and survival of juveniles on patches of coral rubble. J. Exp. Mar. Biol. Ecol. 24, 297–306

Sale, P.F. 1977. Maintenance of high diversity in coral reef fish communities. Am. Nat. 111, 337–359

Sale, P.F. 1978. Reef fishes and other vertebrates: A comparison of social structures. In: Reese, E.S., and Lighter, E.J. (eds.), Contrasts in Behavior. Adaptations in the Aquatic and Terrestrial Environments, Wiley, New York, pp. 313–346

Sale, P.F. 1980. The ecology of fishes on coral reefs. Oceanogr. Mar. Biol. Annu. Rev. 18, 367–421

Sale, P.F. 1985. Patterns of recruitment in coral reef fishes. Fifth Int. Coral Reef Congr., Proc. 5, 391–396

Sale, P.F. 1988. Perception, pattern, chance, and the structure of reef fish communities. Env. Biol. Fish. 21, 3–15

Sale, P.F., Doherty, P.J., Eckert, G.J., Douglas, W.A., and Ferrell, D.J. 1984a. Large scale spatial and temporal variation in recruitment to fish populations on coral reefs. Oecologia 64, 191–198

Sale, P.F., and Douglas, W.A. 1984. Temporal variability in the community structure of fish on coral patch reefs and the relation of community structure to reef structure. Ecology 65, 409–422

Sale, P.F., Douglas, W.A., and Doherty, P.J. 1984b. Choice of microhabitats by coral reef fishes at settlement. Coral Reefs 3, 91–99

Sale, P.F., and Dybdahl, R. 1975. Determinants of community structure for coral reef fishes in an experimental habitat. Ecology 56, 1343–1355

Sale, P.F., and Dybdahl, R. 1978. Determinants of community structure for coral reef fishes in isolated coral heads at lagoonal and reef slope sites. Oecologia 34, 57–74

Sale, P.F., Eckert, G.J., Ferrell, D.J., Fowler, A.J., Jones, T.A., Mapstone, B.D., and Steel, W.J. 1986. Demography of selected aquarium fishes and implications for the management of their collection. Unpub. final report to Great Barrier Reef Marine Park Authority

Sale, P.F., and Ferrell, D.J., in press. Early survivorship of juvenile coral reef fishes. Coral Reefs, in press

Sale, P.F., and Steel, W.J. 1986. Random placement and the structure of reef fish communities. Mar. Ecol. Prog. Ser. 28, 165–174

Sale, P.F., and Williams, D.McB. 1982. The ecology of coral reef fishes: are the patterns more than those expected by chance? Am. Nat. 120, 121–127

Sanders, H.L. 1968. Marine benthic diversity: A comparative study. Am. Nat. 102, 243–282

Sano, M., Shimizu, M., and Nose, Y. 1984a. Food habits of teleostean reef fishes in Okinawa Island, southern Japan. Univ. Mus. Univ. Tokyo, Bull. 25, 1–128

Sano, M., Shimizu, M., and Nose, Y. 1984b. Changes in structure of coral reef fish communities by destruction of hermatypic corals: Observational and experimental views. Pacific Sci. 38, 51–79

Schroeder, R.E. 1985. Recruitment rate patterns of coral reef fishes at Midway lagoon (northwestern Hawaiian Islands). Fifth Int. Coral Reef Congr., Proc. 5, 379–384

Shulman, M.J. 1985. Recruitment of coral reef fishes: effects of distribution of predators and shelter. Ecology 66, 1056–1066

Smith, C.L., and Tyler, J.C. 1972. Space resource sharing in a coral reef fish community. Nat. Hist. Mus. Los Angeles Cty. Sci. Bull. 14, 125–170

Smith, C.L., and Tyler, J.C. 1975. Succession and stability in fish communities of dome-shaped patch reefs in the West Indies. Am. Mus. Novitates 2572, 1–18

Sousa, W.P. 1981. Disturbance in the marine intertidal boulder fields: The nonequilibrium maintenance of species diversity. Ecology 60, 1225–1239

Sousa, W.P. 1984. Intertidal mosaics: Patch size, propagule availability, and spatially variable patterns of succession. Ecology 65, 1918–1935

Stark, W.A., and Schroeder, R.E. 1970. Investigations on the grey snapper, *Lutjanus griseus*. Stud. Trop. Oceanogr. 10, 1–224

Strong, D.R., Simberloff, D.S., Abele, L.G., and Thistle, A. (eds.). 1984. Ecological communities: Conceptual issues and the evidence. Princeton Univ. Press, Princeton, NJ

Sweatman, H.P.A. 1983. Influence of conspecifics on choice of settlement sites by larvae of two pomacentrid reef fishes (*Dascyllus aruanus* and *D. reticulatus*) on coral reefs. Mar. Biol. 75, 225–229

Sweatman, H.P.A. 1985. The influence of adults of some coral reef fishes on larval recruitment. Ecol. Monogr. 55, 469–485

Talbot, F.H., Russell, B.C., and Anderson, G.R.V. 1978. Coral reef fish communities: Unstable, high-diversity systems? Ecol. Monogr. 49, 425–440

Thompson, R., and Munro, J.L. 1978. Aspects of the biology and ecology of Caribbean reef fishes: Serranidae (hinds and groupers). J. Fish. Biol. 12, 115–146

Thresher, R.E. 1984. Reproduction in reef fishes. T.F.H. Publ., Neptune City, NJ

Victor, B.C. 1982. Daily otolith increments and recruitment in two coral reef wrasses, *Thalassoma bifasciatum* and *Halichoeres bivittatus*. Mar. Biol. 71, 203–208

Victor, B.C. 1986. Larval settlement and juvenile mortality in a recruitment-limited coral reef fish population. Ecol. Monogr. 56, 145–160

Waldner, R.E., and Robertson, D.R. 1980. Patterns of habitat partitioning by eight species of territorial Caribbean damselfishes (Pisces: Pomacentridae). Bull. Mar. Sci. 30, 171–186

Walker, M.H. 1978. Food and feeding habits of *Lethrinus chrysostomus* Richardson (Pisces: Perciformes) and other lethrinids on the Great Barrier Reef. Aust. J. Mar. Freshw. Res. 29, 623–630

Walsh, W.J. 1987. Patterns of recruitment and spawning in Hawaiian reef fishes. Env. Biol. Fish. 18, 257–276

Warner, R.R., and Chesson, P.L. 1985. Coexistence mediated by recruitment fluctuations: A field guide to the storage effect. Am. Nat. 125, 769–787

Williams, D.McB. 1980. Dynamics of the pomacentrid community on small patch reefs in One Tree Lagoon (Great Barrier Reef). Bull. Mar. Sci. 30, 159–170

Williams, D.McB. 1983. Daily, monthly and yearly variability in recruitment of a guild of coral reef fishes. Mar. Ecol. Prog. Ser. 10, 231–237

Williams, D.McB. 1986. Temporal variation in the structure of reef slope fish communities (central Great Barrier Reef): short-term effects of *Acanthaster planci* infestation. Mar. Ecol. Prog. Ser. 28, 157–164

Williams, D.McB., and Sale, P.F. 1981. Spatial and temporal patterns of recruitment of juvenile coral reef fishes to coral habitats within One Tree Lagoon, Great Barrier Reef. Mar. Biol. 65, 245–253

Williams, D.McB., Wolanski, E.R., and Andrews, J.C. 1984. Transport mechanisms and the potential movement of planktonic larvae in the central region of the Great Barrier Reef. Coral Reefs 3, 229–236

Woodley, J.D., Chornesky, E.A., Clifford, P.A., Jackson, J.B.C., Kaufman, L.S., Knowlton, N., Lang, J.C., Pearson, M.P., Porter, J.W., Rooney, M.C., Rylaarsdam, K.W., Tunnicliffe, V.J., Wahle, C.M., Wulff, J.L., Curtis, A.S.G., Dallmeyer, M.D., Jupp, B.P., Kochl, M.A.R., Neigel, J., and Sides, E.M. 1981. Hurricane Allen's impact on Jamaican coral reefs. Science 214, 749–755

2. Reef Fish Community Structure: An Indo-Pacific Comparison

Mireille L. Harmelin-Vivien

Coral reefs are well known as the world's most complex marine ecosystems in which fish communities reach their highest degree of diversity. Several, most often contradictory, explanations of this phenomenon have been proposed, mainly based on the degree of predictability and order of these communities (Smith, 1978; Sale, 1978; Bohnsack, 1983). However, species "richness" of reef fish communities also differs according to both geographic areas and types of coral reefs, and it is important to ask why this is so. The present comparative study of reef ichthyofaunas of Tulear, Madagascar and Moorea, French Polynesia, was intended to consider the problem of reef fish community structure slightly differently. The questions we will try to answer can be formulated as follows: Does the difference in species "richness" between the two regions lead to some modifications in the taxonomic and/or functional structure of their fish communities? Do similar functional groups (guilds) of fish exist on both reefs, and what role do they play in the functioning of the ecosystem? Are there identical patterns of resource utilization in time and space? What factors can be proposed to explain the different patterns observed, and at what scale do they operate? In conclusion, the mechanisms involving biotic and abiotic factors that may act on reef fish community structure will be discussed.

The following comparison is mainly based on field work conducted at Tulear, Madagascar, by Vivien (1973), Harmelin-Vivien (1979, 1981), and Maugé (1967) between 1966 and 1972, and at Moorea, French Polynesia by Galzin (1977, 1979, 1985), Bouchon-Navarro (1981), and Harmelin-Vivien and Bouchon-Navarro (1983) since 1977, to whom the reader is referred for further details on the reef studied, site

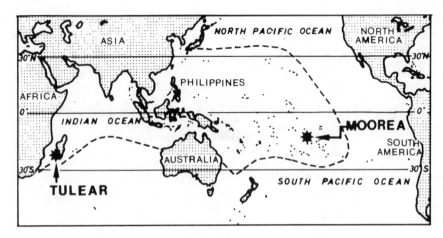

Figure 2.1. The two sites compared, Tulear (Madagascar) and Moorea (French Polynesia), are located at both extremities of the Indo-Pacific region.

description, methodology, and lists of reef fishes. The two study sites are situated at both ends of the Indo-Pacific region. Tulear lies under the Tropic of Capricorn on the southwest coast of Madagascar, in the west Indian Ocean, whereas Moorea is an island in the Society archipelago, French Polynesia, in the central South Pacific Ocean (Figure 2.1). On both islands, the study sites, i.e., the area north of the main barrier reef in front of Tulear, and Tiahura on the northern coast of Moorea, include a reef area a few tens to a few hundreds of meters wide crossing over the reef complex from the shore to the outer slope, and including all the geomorphological zones. Reef fish species were recorded by day and by night in the two sites using the same methods: visual underwater censuses and rotenone sampling.

Results

Total Species Richness

The overall species richness of the coral reef fish faunas studied reaches 552 species from 71 families at Tulear (Harmelin-Vivien, 1979) and 280 species from 48 fami-

Table 2.1. Total Species Richness of Reef Ichthyofauna on Tulear, Madagascar and Moorea, French Polynesia

	Tulear			Moorea		
	Great Reef area	Whole region	Madagascar	Tiahura reef	Whole island	French Polynesia
Number of species	552	820	>1200	280	350	800
Number of families	71	113	>120	48	60	101

Table 2.2. Fish Families and Number of Fish Species per Family Observed on Tulear and Moorea Coral Reefs

Family	Tulear	Moorea	Family	Tulear	Moorea
Carcharhinidae	2	1	Branchiostegidae	1	1
Dasyatidae	2	–	Carangidae	4	2
Torpedinidae	1	–	Lutjanidae	12	4
Rhinobathidae	2	–	Pomadasyidae	5	–
Elopidae	–	1	Lethrinidae	7	3
Moringuidae	5	1	Sparidae	1	–
Muraenidae	37	14	Mullidae	8	8
Congridae	3	–	Pempheridae	2	1
Ophichthidae	15	–	Kyphosidae	1	2
Clupaeidae	1	–	Ephipiidae	1	2
Synodontidae	3	3	Chaetodontidae	21	19
Plotosidae	1	–	Pomacanthidae	6	3
Gobiesocidae	1	–	Pomacentridae	41	25
Antennariidae	2	–	Cirrhitidae	6	8
Ophidiidae	2	1	Mugilidae	–	4
Carapidae	1	–	Sphyraenidae	2	2
Hemirhamphidae	2	–	Polynemidae	–	2
Atherinidae	4	–	Labridae	62	38
Belonidae	–	2	Scaridae	14	12
Holocentridae	15	10	Mugiloididae	2	1
Aulostomidae	1	1	Blenniidae	25	6
Fistulariidae	1	1	Congrogadidae	2	–
Centriscidae	1	–	Tripterygiidae	7	4
Syngnathidae	15	2	Callionymidae	4	–
Scorpaenidae	22	6	Gobiidae	50	11
Caracanthidae	1	–	Microdesmidae	1	–
Synancejidae	2	–	Acanthuridae	18	19
Platycephalidae	4	1	Siganidae	4	2
Serranidae	24	10	Scombridae	1	–
Grammistidae	1	1	Bothidae	2	2
Pseudochromidae	1	1	Pleuronectidae	1	1
Pseudogrammidae	2	1	Soleidae	2	–
Plesiopidae	1	–	Cynoglossidae	2	–
Acanthoclinidae	1	–	Balistidae	17	11
Theraponidae	2	–	Ostraciontidae	4	3
Kuhliidae	–	3	Tetraodontidae	11	9
Priacanthidae	2	1	Diodontidae	2	1
Apogonidae	24	14			

Source: Galzin, 1985 and Harmelin-Vivien, 1979.

lies at Moorea (Galzin, 1985). Thus the fish fauna of the Tulear study area is far richer than that of Moorea (Table 2.1). This high species richness at site level is also found on broader spatial scales (Table 2.1). The Tulear region harbors at least 820 reef fish species, while the whole island of Moorea has only 350 species. On an even broader geographic scale, the Madagascar coral reef ichthyofauna is likely to exceed 1200 species, whereas that of French Polynesia only approximate 800 species (Randall, 1985).

Table 2.3. Reproductive Strategies Based on Egg Types Spawned by Reef Fishes on Tulear and Moorea Reefs

	Tulear		Moorea	
Egg Types	S	%	S	%
Species with pelagic eggs	338	61.2	192	68.6
Species with demersal eggs	205	37.2	86	30.7
Viviparous species	9	1.6	2	0.7

S = number of species; % = percentage in species number.

Taxonomic Structure

The taxonomic composition of the two ichthyofaunas is rather similar at family level, most of the reef fish families displaying an Indo-Pacific or even circumtropical distribution (Table 2.2). The 43 families comon to both provinces represent 60.6% of the families collected at Tulear and 89.6% at Moorea. At the species level, the differences in taxonomic structure are more pronounced: the 136 common species represent only 24.6% of the number of fish species at Tulear and 48.6% at Moorea. The families displaying highest species richness at Tulear belong essentially to the apods (Muraenidae, Congridae, Ophichthidae), Syngnathidae, Scorpaenidae, Serranidae, Apogonidae, Lutjanidae, Pomadasyidae, Pomacentridae, Labridae, Blenniidae, and Gobiidae. Most of them live in close connection with reef constructions and have a rather small home range. On the contrary, at Moorea families with a higher species diversity are generally mid-water families with wide home ranges: Elopidae, Belonidae, Kuhliidae, Mugilidae, Polynemidae.

Reproductive Structure: Egg Type

The different types of eggs laid by fish were examined in relation to the dispersal abilities of the species. Three categories of species were distinguished: (1) those with pelagic (small and numerous) eggs, (2) species with dermersal eggs or "guarded" clutches (larger and less numerous eggs), and (3) the viviparous fish (few embryos per litter).

Table 2.4. Size–Class Structure of Coral Reef Ichthyofauna in Tulear and Moorea

	Tulear		Moorea	
Size–classes (cm)	S	%	S	%
0– 15	178	32.2	81	28.9
15– 30	141	25.5	73	26.1
30– 60	132	23.9	83	29.6
60–120	71	12.9	33	11.8
120–240	28	5.1	9	3.2
>240	2	0.4	1	0.4

Note: Species are ranked into broad size-class according to their maximum total length commonly observed. Size-classes roughly follow a geometric progression as well fit to real observations.
S = number of species; % = percentage in species number.

Table 2.5. Relative Importance of Diet Categories in Fish
Faunas on Tulear and Moorea Reefs

	Tulear		Moorea	
Diet Categories	S	%	S	%
Herbivores	56	10.1	43	15.4
Omnivores	86	15.6	43	15.4
Sessile invertebrate browsers	34	6.2	25	8.9
Zooplankton feeders	57	10.3	27	9.6
Carnivores type A[a]	155	28.1	67	23.9
Carnivores type B[b]	142	25.7	59	21.1
Piscivores	22	4.0	16	5.7

[a] Mainly diurnal carnivores, generally feeding on smaller prey.
[b] Mainly nocturnal carnivores, generally feeding on larger prey.
S = number of species; % = percentage in species number.

Most of the species, almost two-thirds in both regions, have pelagic eggs and therefore large dispersal abilities (Table 2.3). The dominance of pelagic egg fish species is a characteristic of the tropics: the lower the latitude, the higher the proportion of pelagic egg fishes. Fishes laying dermersal eggs represent only one-third of species both at Tulear and at Moorea, and are generally small-sized territorial species. Only vew few fish (1-2% of the total number of species) are viviparous.

Except for the viviparous fish and a few particular species, all coral reef fishes have a pelagic larval stage that lasts from a few days to several months, one month being a common length of time (Williams, 1983; Sale, 1984). Therefore, the dispersal ability of larvae hatching from benthic eggs may be important, although less than for pelagic egg larvae. The long pelagic larval stage of reef fishes has important consequences on species dispersal, recruitment patterns, and reef fish community structure (see Sale, 1984 for a more comprehensive discussion of this topic).

Size–Class Structure

Reef fish communities are characterized by a large species size range, varying from a few centimeters for the smallest species (2-3 cm for some *Eviota*) to more than 4-5 m for the largest ones like some sharks. The size–class structure of the reef ichthyofauna is quite similar in the two reefs studied (Table 2.4). Small-size species (<30 cm) predominate; they represent 58% of the community at Tulear and 55% at Moorea.

Trophic Structure

Diets

The major dietary categories, based on the analysis of gut contents, exhibit the same overall importance in both regions, when all the reef fish communities are concerned (Spearman rank correlation coefficient $rs = 0.94, p < 0.01$); however, a slight difference in percentages exist in the number of species concerned ($t = 2.99, p < 0.05$) (Table 2.5). By decreasing order of importance, the major diet types can be ranked as follows: (1) type A carnivores mainly active by day, preying on small organisms, such

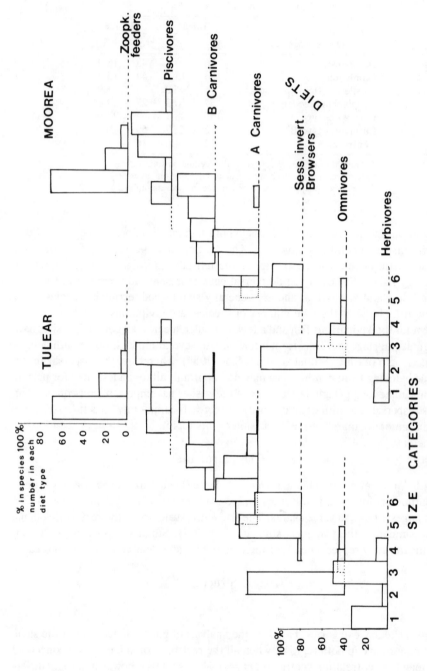

Figure 2.2. Relationships between size and diet in coral reef fishes on Tulear (Madagascar) and Moorea (French Polynesia) reefs. Most omnivores and planktivores are small-size species, whereas some of the carnivores and most piscivores are large-size species.

Table 2.6. Broad Foraging Techniques Among Coral Reef Fishes in Tulear and Moorea

		Tulear		Moorea	
Foraging Techniques		S	%	S	%
Ambushers	Piscivores	9	1.6	4	1.4
	Carnivores (A+B)	47	8.5	16	5.7
Pursuers	Piscivores	13	2.4	12	4.3
and	Carnivores (A+B)	334	60.5	150	53.6
pickers	Zooplankton feeders	57	10.3	28	10.0
	Cleaners	2	0.4	2	0.7
Grazers	Algal grazers	12	2.2	12	4.3
and	Browsers on algal turf	28	5.1	19	6.8
browsers	Browsers on fleshy algae	14	2.5	12	4.3
	Sessile invertebrate browsers	36	6.5	25	8.9

S = number of species; % = percentage in species number.

as crustaceans, polychaetes, molluscs, echinoderms (28% at Tulear, 24% at Moorea); (2) type B carnivores, nocturnal species consuming larger crustaceans and fish (26% at Tulear, 21% at Moorea); (3) omnivores (15% in both reefs); (4) herbivorous species, more numerous at Moorea (15%) than at Tulear (10%); (5) zooplankton feeders (10% at both sites); (6) sessile invertebrate browsers (6% at Tulear, 9% at Moorea); (7) strict piscivorous species with a few species in the two regions (4% at Tulear, 6% at Moorea). The trophic structure of the two reef ichthyofaunas is therefore very similar (Figure 2.2). The main difference lies in the larger number of herbivorous fish species in French Polynesia, which seems to be one of the characteristic features of reef fish communities in this area; this becomes even more obvious when abundances are considered. Total numbers of herbivorous fishes varies from 19 to 55% (mean 33%) on Moorea reefs (Galzin, 1985), and only from 2 to 27% (mean 10%) on Tulear reefs (Harmelin-Vivien, 1979). Conversely, the abundance of carnivorous fish species is greater in Madagascar.

Foraging Techniques

Following Hiatt and Strasburg (1960), Jones (1968), Hobson (1974), and Harmelin-Vivien (1979), three main categories of foraging techniques were distinguished: (1) sit-and-wait predators, (2) pursuing predators, and (3) browsers (Table 2.6). Benthic and mid-water pursuers and pickers are by far the most numerous at Tulear (74%) and at Moorea (69%). On the basis of particular morphological and behavioral adaptations, this latter foraging category could well be further divided into finer groups (Hobson, 1974). Browsers and grazers are well represented into coral reef fish communities, especially in French Polynesia (24% at Moorea, 16% at Tulear). Among herbivores, the relative importances of grazers (22% at Tulear, 28% at Moorea) and browsers (78% at Tulear, 72% at Moorea) is about the same at the two sites. Sit-and-wait predators are comparatively rare in both places (10% at Tulear, 7% at Moorea), but 40% of the piscivorous species hunt by using the sit-and-wait technique at Tulear, and only 25% at Moorea.

Table 2.7. Activity Rhythms of Coral Reef Fishes in Tulear and Moorea

Activity	Tulear		Moorea	
	S	%	S	%
Diurnal species	350	64.4	188	67.4
Nocturnal species	152	27.5	61	21.9
Species active by day and by night	50	9.1	30	10.7

S = number of species; % = percentage in species number.

Diel Temporal Distribution

Most of coral reef fishes exhibit a diel activity rhythm, feeding by day or by night or during both periods. The relative importance of these three activity categories is very similar between Malagasy and Polynesian reef fish faunas (Table 2.7). Diurnally active species predominate. They make up about two-thirds of the community at Tulear (64%) as well as at Moorea (67%). Nocturnal species only make up a quarter of the community and species equally active by day and night are still less numerous (9% at Tulear, 11% at Moorea).

Diurnally and nocturnally active fish assemblages do not only present different species richness; they do not play equivalent roles in the functioning of reef food webs (Harmelin-Vivien, 1979, 1981). The diurnal assemblage exhibits the highest diversity in resource utilization because it includes species belonging to all dietary categories, from herbivores to piscivores. On the contrary, the nocturnal assemblage is trophically less diverse, since it is composed only of carnivorous fishes (zooplankton feeders, type A and B carnivores, piscivores). This diel difference in food resource utilization by coral reef fishes is achieved in the same way on both reefs studied.

Space Distribution

Horizontal Distribution

Horizontal reef fish distribution is not uniform across the reef complex. Its heterogeneity was studied at two different space scales according to (1) geomorphological reef zones characterizing "seascapes" (macrohabitat distribution) and (2) fish habitats (microhabitat distribution).

Seascape Scale. Various reef fish communities may be distinguished on a reef complex; they are generally associated with its main geomorphological reef zones (Goldman and Talbot, 1976; Harmelin-Vivien, 1979; Galzin, 1985, 1987a; Galzin and Legendre, 1988). Richer and more diversified fish communities are found in most Tulear reef zones than in equivalent morphological reef zones at Moorea (Table 2.8). Moreover, the various fish communities studied at Tulear are more different from each other at the species level than their Polynesian counterparts. The percentages of fish species restricted to only one reef zone are higher at Tulear (7–21%, mean 14%) than on the Moorea reef complex (2–12%, mean 5%). The wider range of fish species distribution observed on the Moorea reef is emphasized by the distribution of the species occurrence frequency (Table 2.9). Ubiquitous species are proportionally more numer-

Table 2.8. Fish Species Richness and Relative Importance of Restricted Species in the Various Morphological Reef Zones in Tulear and Moorea

		Tulear		Moorea	
		Total Number Species	% Restricted Species	Total Number Species	% Restricted Species
Littoral seagrass beds		63	20.6	–	–
Fringing reef flat	inner	–	–	54	3.7
	outer	–	–	121	0.0
Inner slope – lagoon		194	7.2	46	0.0
Reef seagrass beds		153	17.6	–	–
Barrier reef flat	inner	92	0.0	–	–
	median	165	18.5	100	1.8
	outer	135	0.0	107	0.0
Boulder tract		131	8.4	–	–
Outer reef flat – reef front		102	8.8	160	3.1
Outer reef slope	spurs-and-grooves	211	15.2	156	12.1
	lower platform	112	18.8	–	–

Type A rare species, abundance <0.1%, excluded from percentages.
– morphological zone absent on the reef studied.

Table 2.9. Occurrence Frequency Distribution of Fish Species in the Different Morphological Reef Zones on Tulear and Moorea Coral Reefs[a]

	Number of Biota in Which Fish Species Are Present							
	1	2	3	4	5	6	7	8
Tulear	43.8	23.9	13.6	7.8	7.1	3.4	0.4	0.0
Moorea	39.3	18.2	12.9	8.2	10.7	8.6	2.1	(–)

(–) = no data.
[a] Expressed as percentages of fish species present in one to n biota.

Table 2.10. Importance of the Various Ecological Categories of Reef Fishes on Tulear and Moorea Reef Tracts[a]

	Tulear		Moorea	
Ecological Categories	S	%	S	%
1 - Species sheltered in reef cavities	69	30.6	58	20.7
2 - Species living in/on sand and rubble	92	16.7	23	8.3
3- Species with small territories (< 1 m)	135	24.5	89	31.8
4 - Swimmers with average home range (> 1 m)	111	20.1	83	29.6
5 - Swimmers with large home range (> 20 m)	27	4.9	8	2.8
6 - Sub-surface species	9	1.6	11	3.9
7 - Pelagic species	9	1.6	8	2.8

S = number of species; % = percentage in species number.
[a] Ecological categories were defined according to diurnal habits of fish species.

ous on Polynesian reefs (11% of species present in six to seven zones at Moorea for 4% at Tulear), whereas species restricted to one or two zones predominate in the Tulear reef ichthyofauna (68%) (Table 2.9).

Habitat Scale. Within a particular geomorphological reef zone, fish distribution at the coral patch scale is highly heterogeneous (within-habitat diversity) (Sale and Dybdahl, 1975; Gladfelter and Gladfelter, 1978; Gladfelter et al., 1980). This heterogeneity may be in part related to specific habitat characteristics of fishes, such as the nature and size of their home range. The larger part, however, is due to the stochastic processes of fish larvae settlement as shown by Sale and others (Sale, 1984; Sale, Douglas and Doherty, 1984; Williams, 1983; Shulman, 1983). To consider this problem, seven ecological categories were distinguished according to the nature of their habitat during daytime and to the amplitude of fish movements (Table 2.10), as follows: (1) species sheltered in reef holes, crevices, or under overhangs during daytime; (2) species living in or on sand and rubble. (3) species defending small territories around coral heads; (4) species swimming above corals in a few square meters; (5) good nectobenthic swimming species with large home ranges; (6) subsurface species; (7) pelagic species. Species included within categories 1–3 generally exhibit a smaller range of displacement than fishes ascribed to categories 4–7.

On both reefs studied, species with small-size home ranges predominate (72% at Tulear, 61% at Moorea). The predominance of fishes with small home ranges at Tulear is mainly due to a higher species richness of muraenids, ophichthids, syngnathids, scorpaenids, serranids, apogonids, blenniids, gobiids, pomacentrids, and labrids. On the contrary, reef and pelagic fishes with large home ranges are less numerous (28% at Tulear, 39% at Moorea), but subsurface and mid-water pelagic fishes are better represented on Polynesian reefs (Table 2.10) suggesting a stronger oceanic influence in this region. A far greater number of ecological categories could well have been distinguished if the composition of specific microhabitats had been taken into account, along with the nocturnal changes of microhabitat displayed by most fish species.

Vertical Distribution

Fishes, like birds, move in a three-dimensional space. The vertical distribution of fishes into reef constructions and surrounding waters is an important and complex factor to take into consideration if one wants to understand reef fish community structure (Harmelin-Vivien, 1979). Broadly speaking, three main categories of vertical space use may be distinguished, which bring to the fore the major differences between Tulear and Moorea fish faunas (Table 2.11): (1) bottom-related species; (2) species moving alternatively between bottom constructions and mid-water layers on a diel cycle (this category may be subdivided into two groups: species that swim close to the corals and those that enter mid-water to feed); (3) permanent mid-water species.

Species that change vertical layers on a diel cycle are the most numerous on both reefs (Table 2.11), especially at Moorea (76%). Mid-water species are the less abundant (at Moorea 7%, at Tulear 3%). The main difference between the two reefs is the bottom-related species, which represent nearly 40% of the community at Tulear and only 18% at Moorea.

Table 2.11. Broad Vertical Space Utilization Categories of Coral Reef Fishes in Tulear and Moorea

| | | Tulear | | Moorea | |
		S	%	S	%
Bottom-related species		218	39.5	49	17.5
Species using coral constructions		316	57.3	212	75.7
and above water layer	(Type A)	(259	47.0)	(185	66.1)
	(Type B)	(57	10.3)	(27	9.6)
Mid-water species		18	3.2	19	6.8

S = number of species; % = percentage in species numbers.
(Type A) = species swimming in the coral proximate water layer.
(Type B) = species entering mid-water to feed.

At the small coral patch scale, the reef represents a tridimensional mosaic. Its heterogeneity differs according to the body size and the home range of the species, which influence the ways they can make use of the different resources. In such a tridimensional mosaic, resource sharing varies among trophic, ecological, and size categories on a day–night cycle, and in vertical as well as horizontal directions.

Table 2.12. Distribution of Coral Reef Fish Species Among Both Trophic and Ecological Categories on Tulear (A) and Moorea (B) Reefs[a]

(A) Tulear

| Diets | Ecological Categories | | | | | | | Total Species Number |
	1	2	3	4	5	6	7	
Herbivores	3.3	–	0.2	6.7	–	–	–	56
Omnivores	0.2	7.3	8.0	–	0.2	–	–	86
Ses. inv. browsers	–	–	6.2	–	–	–	–	34
Zooplankton feeders	6.0	–	2.2	0.5	0.7	–	0.9	57
Carnivores type A	6.2	6.0	1.6	12.9	1.1	–	0.4	155
Carnivores type B	15.0	1.8	6.3	–	2.5	–	–	142
Piscivores	–	1.6	–	–	0.4	1.6	0.4	22
Total species number	169	92	135	111	27	9	9	552

(B) Moorea

| Diets | Ecological Categories | | | | | | | Total Species Number |
	1	2	3	4	5	6	7	
Herbivores	2.1	–	0.7	12.5	–	–	–	43
Omnivores	–	2.9	10.0	–	–	–	2.5	43
Ses. inv. browsers	–	–	8.6	0.4	–	–	–	25
Zooplankton feeders	6.4	0.4	2.5	0.4	–	–	–	27
Carnivores type A	1.4	2.5	2.5	16.4	0.4	0.7	–	67
Carnivores type B	10.7	1.1	7.5	–	1.8	–	–	59
Piscivores	–	1.4	–	–	0.7	2.1	1.4	16
Total species number	58	23	89	83	8	8	11	280

[a]Expressed as percentages in species number.

Food and Space Utilization

How do the numerous coral reef fish species share food and space resources? First, it must be noticed that only a few functional groups (guilds) actually exist among all those possible in theory if one combines the two parameters (Table 2.12). As a whole, the patterns of food and space use are very similar on both reefs, most functional groups (guilds) occurring at Tulear as well as at Moorea (Figure 2.3). Those present in only one reef are of minor importance. Nevertheless, the greater importance of some guilds in one or the other reef indicates a slightly different use of foraging zones by the various trophic categories. Among the herbivores, small-sized species sheltered into reef holes (H1 = mainly blenniids) are more numerous at Tulear, whereas large swimming species such as scarids, acanthurids, and siganids (H4) are more important at Moorea (Figure 2.3). Among the omnivores, sandbottom species like gobiids (O2) are better represented on the Tulear reef. All ecological categories of sessile invertebrate browsers are more numerous at Moorea. Among the zooplankton feeders, two peaks of importance are equally observed in the two regions: one due to fish species living in reef holes by day and entering mid-water at night to feed (Z1 = apogonids, holocentrids, pempherids . . .), the other due to diurnal territorial species, mainly pomacentrids (Z3). The occurrence of mid-water plankton feeders at Tulear (Z6, Z7) is due to schools of clupeids and atherinids. Among the carnivores, species living in reef cavities during at least one period (CA1, CB1) are proportionally more numerous at Tulear, whereas species always swimming above or around corals (CA4, CB3) are more important at Moorea. The distribution of fish-eating species among the various ecological categories is rather similar in the two regions, with a slightly higher proportion of pelagic piscivores (P7) at Moorea.

Although the patterns of food and space utilization are similar on both reefs, a better and more diversified utilization of reef cavities is observed on Tulear reef.

Fish Community Structure and Reef Morphology

Until now the comparison of the Malagasy and Polynesian fish communities was essentially conducted at the scale of a whole reef. However, on a smaller scale the fish community structure of a reef flat is very different from that of an outer reef slope. The division of a reef complex into these two broad morphological zones separates two groups of biotopes in which the range of variation of environmental factors differs greatly. Temperature, salinity, oxygen, light, water level, and currents vary much more on reef flats than on the outer slope, and may influence fish community structure differently (Figure 2.4).

(1) Activity rhythms: No change in community structure is noticeable; the proportions of diurnal and nocturnal fish species are similar on the reef flat and on the outer slope at Tulear as at Moorea. (2) Egg types: Species with demersal eggs are better represented on the reef flats, and pelagic eggs species on the outer slopes in both areas. (3) Size–class categories: Small-size species (< 15 cm) are more numerous on the reef flats, especially at Tulear. (4) Ecological categories: Species living sheltered during the day in reef cavities, or on sandy bottom, are more numerous on the reef flats, whereas good swimmers predominate on the outer slopes in both areas. (5) Trophic categories: On both reefs, there is an increase in omnivores on the reef flats, and in zooplankton

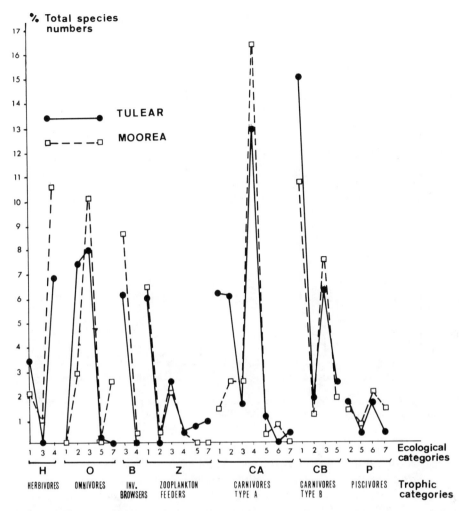

Figure 2.3. Food and space utilization in reef fish communities on Tulear and Moorea coral reefs. Relative importance of fish species according to their distribution into ecological (1,2, . . .) and trophic (H,O, . . .) categories.

feeders on the outer slopes. An increase in sessile invertebrate browsers and piscivores on the outer slope, and in type A carnivores on the reef flat, was noted at Tulear. At Moorea, piscivores are more numerous on the reef flat, and type B carnivores on the outer slope (Figure 2.4). Therefore, the general patterns of fish community structure undergo similar changes on reef flats and outer slopes on both reefs, but the differences between reef flat and outer slope communities are more pronounced on the Tulear reef (Table 2.13).

The differences in the overall functional structure of the ichthyofauna are smaller between the two reefs when they are considered as a whole than between analogous

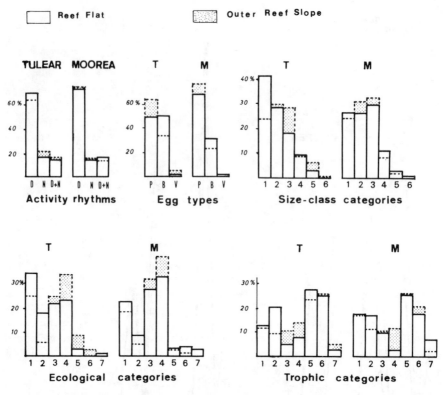

Figure 2.4. Comparative evolution of fish community structure on reef flat and outer reef slope at Tulear and at Moorea (percentages in species number).

geomorphological structures of these same reefs (Table 2.14). Conversely, on a smaller scale, the differences between community structures are increased; these differences would increase even more if the scale were still smaller. For example, the relative proportions of different diets largely differ from one biotope to another across the same reef (Goldman, 1973; Harmelin-Vivien, 1979; Molina, 1983; Galzin, 1985). These differences in community structure are more important between reef flat communities

Table 2.13. Differences in Reef Fish Community Structure on Reef Flat and Outer Reef Slope in Tulear and Moorea Coral Reefs[a]

	Activity Rhythm	Egg Type	Size–Class Category	Ecological Category	Trophic Category
Tulear	7.8	32.8	34.6	42.2	30.4
Moorea	1.2	15.8	14.1	24.8	23.9

[a] Expressed as the total amount of deviation between reef flat and reef slope values for each community parameter ($D = \sum_{1}^{n} |x_{i_T} - x_{i_M}|$, where n = number of class for each parameter; x_{i_T} = value of parameter i in Tulear fish fauna; x_{i_M} = value of parameter i in Moorea fish fauna).

Table 2.14. Mean Differences in Reef Fish Community Structure According to Geographic Regions and Morphological Reef Structures[a]

		Mean + 1 SD
Total difference between Tulear and Moorea		19.54 + 12.30
Difference between reef flat and outer slope on the same reef	Tulear	29.56 + 12.94
	Moorea	15.96 + 9.52
Difference between Tulear and Moorea for the same morphological zone	Reef flat	30.10 + 13.20
	Outer slope	20.26 + 6.13

[a] Mean of total deviations between reef flat and outer reef slope values of the five parameters studied.

than between outer slope communities (Table 2.14). Peyrot-Clausade (1976, 1977a, 1977b) also noticed that the differences between motile cryptofaunas at Tulear and Moorea were greater between reef flat communities than between outer reef slopes.

The Heterogeneity of Fish Distribution and the Problem of the Rare Species

Small-scale studies emphasize the heterogeneity and overdispersion of the coral reef fish distribution (Sale, 1980, 1984; Russell et al., 1974). The distribution of abundance class frequency, based on the percentage occurrence of species in each community, shows that for 85% of the species, each one represent less than 1% of the total number of individuals belonging to this community, at Tulear as well as at Moorea (Table 2.15). A large proportion of rare species was observed in all the biotopes we examined, as shown by the low standard deviation and the low range values of the first (< 1%) abundance class (Table 2.15). Species whose only one specimen was collected or observed in each biotope, and corresponding each to less than 0.1% of the total number of individuals, make up nearly a quarter of the community: 25% at Tulear, 22% at Moorea (Table 2.16)! But the importance of scarce species depends on the definition of scarcity. Thus the proportion of rare species is increased almost by a factor of 2 when the following fish categories are also considered as rare: (1) the species represented by one specimen on the whole reef tract and (2) the species represented by one specimen in each

Table 2.15. Distribution of Abundance-Class Frequency of Fishes on Tulear and Moorea Coral Reefs[a]

Abundance Classes	Tulear ($N = 8$)[b]		Moorea ($N = 3$)[c]	
	Mean	Range	Mean	Range
$0 < n \leq 1\%$	84.9	(71–90)	84.5	(79–90)
$1 < n \leq 5\%$	11.5	(7–18)	10.4	(6–15)
$5 < n \leq 10\%$	3.4	(0– 8)	3.2	(1– 4)
$10 < n \leq 20\%$	1.0	(0– 5)	1.1	(0– 2)
$20 < n \leq 50\%$	0.2	(0– 2)	0.8	(0– 1)

N = number of biota.
[a] As range of species abundance varies within and between reef communities, abundance classes were defined according to the numerical percentage of species.
[b] Data from Harmelin-Vivien, 1979, Table 1.
[c] Data from Galzin, 1985, Tables 8–10.

Table 2.16. Relative Importance of Rare Species in the Fish
Faunas of Tulear and Moorea Coral Reefs According to the
Definition of Species Scarcity[a]

Rareness Status	Tulear[b]	Moorea[c]
Species A	17.0	13.6
Species B	24.7	22.0

[a] Expressed as species numerical percentages. A = species
observed in only one specimen on the whole reef area studied; B =
species observed in one specimen per biota.
[b] Data from Harmelin-Vivien, 1979, Table 1.
[c] Data from Galzin, 1985, Tables 8–10.

biotope (Table 2.16). Rare species related to definition 1 form 17% of species at Tulear
and 13% at Moorea! Most of them are benthic fishes closely dependent on coral constructions, most of them territorial or with a small home range (Table 2.17). They are
not food specialists and can belong to all the trophic categories previously defined with
percentages similar to those computed for the whole fish community in each area (Table
2.18).

Conclusion on Tulear–Moorea Data Comparison

In spite of the existing differences in species richness and taxonomic composition, the
reef ichthyofaunas of Tulear and Moorea share similar functional characteristics: (1) a
dominance of species with pelagic eggs (two-thirds of the species); (2) a wide range of
size classes (from a few centimeters to several meters) with preponderance of small-
size species (LT < 30 cm); (3) a predominance of carnivorous fishes feeding on motile
benthic invertebrates (40–50%), along with rather high proportions of omnivorous and
herbivorous species (15% each); (4) a predominance of fish actively hunting their prey
(70%), browsers and sit-and-wait predators being much less numerous (20 and 10%,
respectively); (5) a majority of diurnal species (two-thirds of species); (6) a change of
habitat and/or of vertical stratification on a diel rhythm in most species; (7) the large
number of rare species (85% of the species of the community represent less than 1% of
the total fish population, and 25% less than 0.1%).

The changes in fish community structure along broad geomorphological zones
(reef flat vs. outer reef slope) follow a same overall trend on Tulear and Moorea reefs.
On the reef flats, a higher proportion of species with dermersal eggs, of small-size

Table 2.17. Distribution of Rare Fish Species into Ecological Categories on Tulear and Moorea
Reefs[a]

	Ecological Categories						
	1	2	3	4	5	6	7
Tulear	34.0	31.9	21.3	6.4	1.1	4.3	1.1
Moorea	15.8	5.3	44.7	23.7	5.3	2.6	5.3

[a] Expressed as percentages in species number.

Table 2.18. Relative Importance of Diet Categories Among Rare Fish Species in Tulear and Moorea Coral Reefs[a]

	Diet						
	Herbiv.	Omniv.	Ses. inv.	Zooplank.	Carniv. A	Carniv. B	Pisciv.
Tulear	8.5	11.7	4.3	7.4	37.2	25.5	5.3
Moorea	23.7	13.2	10.5	0.0	15.8	26.3	10.5

[a] Percentages in species number.

species, of species sheltering in reef holes or living in/on sandy bottoms, and of omnivorous species is noted. Conversely, on outer slopes, an increase in species with pelagic eggs, of large-size species, of species with large home ranges, and of zooplankton feeders is observed.

Nevertheless, from a functional standpoint, some differences exist between the fish communities of Tulear and Moorea; they are mainly due to a change in proportions of some functional groups. These differences may be summarized as follows (+ = higher; − = lower importance):

	Tulear	Moorea
Total species richness	+	−
Species richness per biota	+	−
Characteristic species per biota	+	−
Percentage of restricted species	+	−
Percentage of ubiquitous species	−	+
Importance of carnivores	+	−
Importance of herbivores	−	+
Importance of pursuers	+	−
Importance of browers	−	+
Importance of species sheltered in reef holes	+	−
Importance of species living in sandy areas	+	−
Differences between reef flat and outer slope	+	−

From a structural standpoint, the two-reef ichthyofaunas are similar in both areas. The increased species richness of the Tulear fish community is due to an increase in the number of species belonging to all functional groups, and not to the presence of additional guilds (Figure 2.5). Nevertheless some differences do occur, and all emphasize the larger importance of species strongly dependent on the reef itself. A better use by fishes of all habitat and food resources of the lower and inner layers of the reef tract is a characteristic of Tulear reefs, more particularly on the reef flat structures.

Determinants of Differences in Community Structure

The main factors involved in the evolution and maintenance of reef fish community structure are of historical, biogeographic, geomorphological, architectural, and biotic nature. Let us see how they might contribute to explaining the differences observed in

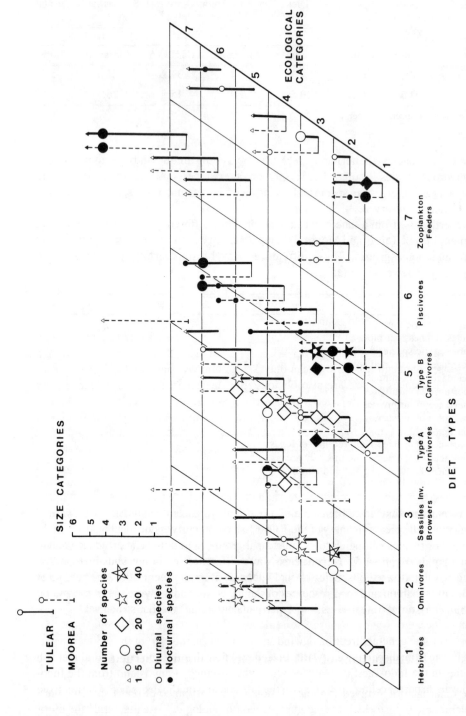

Figure 2.5. Global fish community structure on Tulear and Moorea reefs: size, diets, microhabitat utilization (ecological categories), and diel activity of reef fishes.

Table 2.19. Main Geographic Features of Madagascar and Moorea Islands and Hydrological Characteristics of Reef Surrounding Waters on Tulear and Moorea Coral Reefs

	Tulear	Moorea
Latitude	23°25 S	17°30 S
Longitude	43°40 E	149°50 W
Island surface (km²)	587,000	136
Age of island (million years)	> 200	1.2
Tide range (m)	0.8–2.4(max3.3)	0.1–0.4(max0.5)
Maximum temperature range between day and night on reef tract (°C)	10–20	6
Mean temperature of ocean surface water (°C)	23–30	24–31
Salinity of ocean surface water	34–35.5	35–36
Chlorophylle-a concentration of surface water (mg/m³)	0.20–0.40	0.03–0.05

species richness and community structure between Tulear and Moorea reef ichthyofaunas, and at what temporal or spatial scales they might be operating.

Historical and Biogeographical Factors

Madagascar is a very old island that was separated from Africa during the Jurassic period (140 million years) when the continent of Gondwana was split up into several continental plates. The island of Moorea is much younger, having been formed from a hot spot in the central Pacific Ocean only 1.2 million years ago (Table 2.19).

The differences in size are also obvious: the surface area of Madagascar reaches 578,000 km² and greatly exceeds that of Moorea (132 km²). These two islands are located at the two extremities of the Indo-Pacific region: Madagascar in the southwest at latitude 23° south and Moorea in the central-east at latitude 17° south (Figure 2.1). Both are equally far from the Indo-Malayan archipelago, which is presumed to be the dispersal center of all coral reef fishes. But Madagascar is located close to the African continent and is connected to Southeast Asia by numerous coral islands (Nicobars, Maldives, Laccadives, Seychelles, etc.) and the reefs scattered along the coast of India, Sri Lanka, the Arabian peninsula, and East Africa. In general winds blow westward in the northern tropical zone of the Indian Ocean, and the currents flow in the same direction, dispersing reef fish larvae toward the western shores of the Indian Ocean. Moreover, oceanic waters surrounding Madagascar are rich in nutrients provided by telluric waters from the island itself and by the African continent (Sournia, 1972). On the contrary, Moorea, like all Polynesian islands, is isolated in the center of the Pacific Ocean, far from the Australian and American continents, and is surrounded by oligotrophic oceanic waters (Sournia and Ricard, 1976; Gabrié and Salvat, 1985). Moreover, the general direction of winds and currents in this part of the world prevents fish larvae from dispersing eastward. Indeed we know that the number of coral reef fish species in the Indo-Pacific area depends on the distance of any island from the Indo-Malayan dispersal center, as well as on the latitude and on the degree of isolation from continental masses (Allen, 1975). The first two factors cannot be put forward to explain an increased reef fish species number in Madagascar reefs, but the proximity of the

Table 2.20. Relation Between Species Richness of Corals and Reef Fishes on a Biogeographic Scale

Locality	Ref.	Number of Coral spp.	Number of Fish spp.
Koweit	Downing, 1985	23	85
Djibouti, Tadjoura	Laborel, in prep.; Saldanha, in prep.	65	180
Tutia Reef	Talbot, 1965	52	192
La Reunion			
Baie Possession	Bouchon & Bouchon-Navarro, 1981	54	109
Hermitage	Bouchon & Bouchon-Navarro, 1981	30	81
St. Gilles	Bouchon, 1981 Harmelin-Vivien, 1976	120	258
Aqaba	Bouchon, in prep. Bouchon-Navarro, in prep.	150	400
Tulear	Pichon, 1978 Harmelin-Vivien, 1979	147	552
Moorea	Bouchon, 1985; Galzin, 1985	48	280
Society Islands	Pichon, 1985; Galzin, 1985	120	633
French Polynesia	Pichon, 1985; Randall, 1985	168	800
Heron Island	Mather & Bennet eds, 1984	139	750
New Caledonia	Robin et al., 1980; Fourmanoir & Laboute, 1976	300	1000
Great Barrier Reef	Reeder's Digest ed, 1984	500	2000

Note: Numbers of coral and fish species are given for similar reef surfaces in each locality, but size of reef area surveyed differs between localities. Linear regression curve: $y = -13.63 + 3.92x$, $r = 0.97$, $p < 0.01$.

African continent, the effect of winds and currents on reef fish larval dispersal, and the high nutrient supply of neritic waters all must contribute to this increase. In the Indo-Pacific region, therefore, biogeographic factors seem to prevail over historical ones as a result of the absence of discontinuity in the carrier element (the intertropical waters) between sites, and the duration of the larval period of these fishes, which increases their potential for long-range dispersal.

Geomorphological Factors

The structure and extent of geomorphological zones differ in the two reefs studied, the Tulear reef being the wider and more diverse. The barrier reef flat is 1500 m wide at Tulear and only 490 m at Moorea. The inner and outer reef flats and the outer spur and groove system are structurally more complex at Tulear. Some geomorphological zones, such as the boulder tract, the seagrass beds and the deep outer flagstone, which all harbor highly characterized fish assemblages at Tulear, do not exist at Moorea. The more complex geomorphological structure of the reef no doubt contributes to the increase in fish species richness at Tulear.

Architectural and Biotic Factors

As in a tropical rainforest, the most distinctive architectural features of a coral reef are of biotic origin, but trees are here replaced by sessile invertebrates, the scleractinians, which likewise contribute to an important part of the reef primary production through

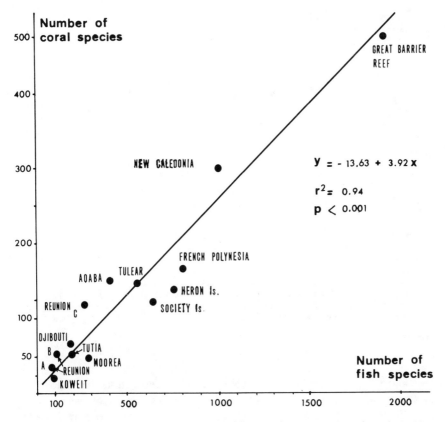

Figure 2.6. Linear regression between number of fish species and number of coral species on a biogeographic Indo-Pacific scale.

their numerous symbiotic algae. Corals are the major reef builders, except on algal ridges where calcareous algae become the principal architectural components. The high diversity of tropical land plants is often considered a likely cause of the high species richness of rainforest vertebrate communities. But what about coral–fish and algae–fish relations on coral reefs?

The scleractinians made their appearance during the Triassic period, and expanded in number and variety since the Jurassic. Teleostean radiation began during the Jurassic and the fish evolution was largely concomitant with that of coral reefs (Gosline, 1971). On a larger biogeographical scale, a strong positive correlation also exists between coral and fish species richness (Table 2.20, Figure 2.6). However, on a smaller scale, on the various geomorphological reef zones, such a relationship is far less obvious and these two factors are not strongly correlated. On the Tulear reef as a whole, a slight positive correlation exists between coral and fish species richness. In fact, a strong positive correlation ($p < 0.01$) is also observed on the outer reef slope but no longer exists on the reef flat (Table 2.21). Conversely, at Moorea no correlation exists on the outer slope and reef flat formations. Furthermore, at Tulear as well as at Moorea, fish species richness is never correlated with overall algal species richness (Table 2.21).

Table 2.21. Species Richness Correlation in Coral and Fish Communities and in Algae and Fish Communities on Tulear and Moorea Reefs

	Tulear[a]			Moorea[b]		
	Total Reef $n = 13$	Slopes $n = 7$	Reef Flat $n = 6$	Total Reef $= 9$	Slopes $n = 4$	Reef Flat $n = 5$
Coral species richness vs. fish species richness	$r = 0.67$ $p < 0.05$	$r = 0.91$ $p < 0.01$	$r = 0.10$ n.s.	$r = 0.32$ n.s.	$r = 0.51$ n.s.	$r = 0.32$ n.s.
Percentage of live coral coverage vs. fish species richness	$r = 0.14$ n.s.	$r = 0.76$ $p < 0.05$	$r = 0.17$ n.s.	$r = 0.53$ n.s.	$r = 0.63$ n.s.	$r = 0.52$ n.s.
Algae species richness vs. fish species richness	$r = 0.36$ n.s.	$r = 0.26$ n.s.	$r = 0.26$ n.s.	–	–	$r = 0.14$ n.s.

[a] Data from Pichon (1978) for corals and algae and from Harmelin-Vivien (1979) for fishes.
[b] Data from Bouchon (1985) for corals, from Payri (1982) for algae, and from Galzin (1985) for fishes.
r = linear regression coefficient; n = number of data set; p = level of statistical significance; n.s. = no significance.

Some authors, however, noticed a positive correlation between the percentage of live coral cover and the species richness and abundance of coral reef fishes (Bell and Galzin, 1984). Furthermore, it has also been noticed that the presence of some fishes with specific diets ("specialists") depend on the presence of specific coral growth forms. Bouchon-Navarro et al. (1985), for example, definitely established that the abundance of exclusive coral feeders within the Chaetodontidae is positively correlated with the abundance of tall-branched coral colonies. It can therefore be concluded that a positive correlation between coral and coral reef fish diversities only exists on a large (biogeographic) scale; it no longer persists on a smaller (within-reef) scale, except in a very few cases.

Whereas studies on coral "architectural models" have just begun (Dauget, 1986), those concerning the architecture of whole coral communities remain to be undertaken. The matter should not be overlooked because it has been shown that, in certain terrestrial habitats at least, foliage diversity is a better predictor of bird species richness than tree species richness itself (McArthur, 1972; Pearson, 1977). For reef fishes, Risk (1972) and Luckhurst and Luckhurst (1978) also showed that species richness is positively correlated to reef "complexity." As the one coral species can present different growth forms under different environmental conditions, the species composition of a given coral community cannot tell us much about its architecture—the less so since identical architectural models can occur, for instance, at Tulear and at Moorea (Pichon, 1978; Bouchon, 1985). Nevertheless, Peyrot-Clausade (1976, 1977a, 1977b), studying motile cryptofauna on the two reefs considered, noticed that the development and heterogeneity of reef cavities were higher at Tulear than at Moorea particularly in reef flat formations, the differences in reef internal structure being smaller on outer reef slopes.

All the factors previously considered—island age and size, distance from continents, nutrient contents of waters, geomorphological reef zone diversity, and coral species richness—cannot but contribute to a higher fish species richness on the Tulear reef. The major differences between Tulear and Moorea are more to be found between the reef flat formations of the two reefs than between their outer slopes. It is also between the reef flat fish communities that species richness and functional structure differ the most. If biogeographic factors seem to be mainly responsible for the overall difference in total species richness, geomorphological, biotic, and architectural factors appear to account for most of the differences in diversity and functional structure of fish communities at the smaller (within-reef) scale.

Discussion

Maintenance of High Species Richness on Coral Reefs

The high species richness of tropical communities has long aroused the interest of ecologists (cf. Pianka, 1966, 1974; McArthur, 1972; Whittaker, 1977; Ricklefs, 1979; Bourlière, 1983; and many others), and various hypotheses were put forward to explain such a high diversity (cf. reviews of Rohde, 1978 and Huston, 1979). Possible causes and mechanisms for maintaining high species richness in tropical communities, particularly the respective role of biotic interactions and abiotic factors on coexistence patterns, dispersion processes, and extinction rates of species, gave rise to an abundant, often controversial literature (for general reviews, see, for example, McArthur and Wilson, 1967; Cody and Diamond, 1975; Brown, 1981; Strong et al., 1984). This problem is also of direct concern for those involved in studies of reef fish community structure. The alternative, non-mutually exclusive theories proposed to account for the coexistence of a great number of fish species on coral reefs were well summarized and discussed by Sale (1977, 1980, 1984), Talbot et al. (1978), Anderson et al. (1981), Bohnsack (1983), and Abrams (1984). Although they differ in the importance they give to the different factors in the structuration of reef fish communities, most of them agree that the following factors definitely contribute to the high species richness of coral reef fish communities: the spatial and temporal resource heterogeneity, the differential temporal use of resources by co-occurring species, and some biological interactions taking place between species (competition, predation, and mutalism). I do not intend to review this problem once more but want only to comment on some points on the basis of my own observations.

First, it has long been postulated that tropical communities were "saturated" and "closed" (Elton, 1958; Hutchinson, 1959; McArthur and Wilson, 1967; and others), all ecological niches being filled and their spatial dimensions being narrower than in temperate latitudes. Price (1980, 1984) and Lawton (1982, 1984) provided a number of examples of nonsaturated communities, even in the tropics, as well as of vacant niches in specialist insect and parasite communities. In spite of the high species richness of fish and invertebrate communities in coral reefs, there is, to my knowledge, no evidence that space is a priori always saturated, in numbers of species as well as in individuals. Abele (1984) showed that it is experimentally possible to generate artificial oversaturated communities of coral-associated crustaceans. Sale (1972) found no correlation between the abundance of the damselfish *Dascyllus aruanus* and the size of coral colonies in which they lived when coral colonies are very abundant. The temporal variations of fish species and individual numbers on a given transect at various time scales tend also to suggest that space, at least at certain periods of the year, is not a limiting factor (Kock, 1982; Molina, 1983; Galzin, 1985, 1987b; Williams, 1986). If some fish species exhibit asynchronous density variations which may be interpreted as a kind of temporal space partitioning (Kock, 1982; Molina, 1983), most generally they increase their numbers and/or their abundances synchronously (Kock, 1982; Molina, 1983; Galzin, 1985, 1987b; Williams, 1983, 1986). Reactions of reef fish communities to natural catastrophes also indicate that space may not be a priori saturated in the reef environment. Underwater visual counts of outer slope fish communities, before and after total destruction of coral constructions by cyclones in several French Polynesian islands, have shown that (1) as expected, species richness and abundance of fishes decreased in areas destroyed by storms, (2) they did not change in uninjured areas located far from those affected, and (3) both diversity and abundance of fish greatly increased in the intact areas close to the perturbed ones (Table 2.22). The fish community increase in the latter areas was not an ephemeral phenomenon and persisted over more than 6 months after the cyclone strike. On space utilization by reef fishes, Sale (1984) concluded that "such changes in numbers of fish present indicate that space resources are either not fully used for much of the time, or that space requirements of particular species vary dramatically from time to time" (p. 483). This is indeed the case during the ontogenic development of species or during the breeding season.

Spatial and Temporal Heterogeneity of Resources

Coral reefs are well known to represent a patchy environment model, with patchiness occurring in both the horizontal and vertical directions (different types of substrata with various rugosity, different coral species with different growth forms, and so on). Therefore a coral reef is a tridimensional mosaic of microhabitats, randomly distributed at a patch-reef scale but which form relatively homogeneous, characteristic "seascapes" at a larger scale. At the scale of a whole reef, a striking environmental heterogeneity is again evident due to structural differences between various seascapes.

Coral reefs were also considered for a long time as stable, well-ordered ecosystems at or near equilibrium, a viewpoint strongly questioned nowadays (cf. Connell and Orias, 1964; Connell, 1978, 1983; Sale, 1977, 1980, 1984). The nature and structure of the reef tridimensional mosaic varies in time in relation with more or less rhythmic short-term changes and also with aperiodic events, whose effects range from the within-colony level to the whole region. Research on temporal modifications of the coral reef communities began only recently. Variations in growth rate, calcification, morphology, reproduction, and productivity were evidenced for the main architectural reef components: scleractinians (cf. Harrison et al., 1984; Schlesinger and Loya, 1985; Loya, 1985; and others) and algae (cf. Payri, 1982; Morrissey, 1985). Taking into account their heterogeneity in space and time, coral reefs can presently be considered as a benign but uniformly variable environment (Sale, 1977; Comins and Noble, 1985), subjected to disturbances of intermediate intensity and frequency (Connell, 1978). Nevertheless, intense and rare disturbances such as cyclones, *Acanthaster* proliferation, or various epidemics deeply influence reef community structure and their

Table 2.22. Increase in Density of Herbivorous Fish Assemblage on Outer Reef Slope of Takapoto Atoll (Tuamotu Archipelago) in a Reef Zone Located in the Proximity of a Reef Sector Totally Destroyed by Hurricanes[a]

	Takapoto		Moorea	
	Oct. 1982 (before hurricanes)	Oct. 1983 (after hurricanes)	Oct. 1982 (before hurricanes)	Oct. 1983 (after hurricanes
Mean number of fish individuals/250 m²	143.5	323.5	371.8	382.0
+ 1 SD	18.1	37.2	157.3	150.8
T-test level significance	$p < 0.01$		n.s.	

[a] In comparison, data are given for a reef zone (Moorea) located far from hurricane impacts.

associated fish communities by increasing the natural reef heterogeneity on a still larger space and time scale.

An increase in spatial heterogeneity generally produces an increase in species richness of fish communities (Risk, 1972; Luckhurst and Luckhurst, 1978), and different fish assemblages are associated with the various morphological zones of the reef (for discussion, see Talbot and Goldman, 1972; Galzin, 1985, 1987a; Galzin and Legendre, 1988). Temporal variations in reef fish species richness and abundance also occur following cyclic rhythms (Molles, 1978; Molina, 1983; Galzin, 1985, 1987b; Williams, 1986) or stochastic patterns (Sale and Douglas, 1984), but none of them can, for the time being, be correlated with any modification of the reef environment. However, it has been noticed that a decrease in live coral cover can induce modifications of chaetodontid fish assemblages (Bouchon-Navarro et al., 1985; Williams, 1986), although none in the other syntopic fish families (Williams, 1986). It will be necessary to greatly increase our knowledge of the specific needs of the different fish groups if we want to better understand the detailed functioning of fish communities.

Time Partitioning

One of the simplest ways of maintaining a high species richness in any habitat is the time partitioning of space and food resources, and this is what happens in reef fish communities as well as in higher vertebrate communities of rainforests (Charles-Dominique, 1975; Bourlière, 1983). Studies on food webs carried out at Tulear and Moorea reefs emphasized the existence of distinct diurnal and nocturnal fish assemblages, differing in their species richness, their abundance, and their trophic structure. The diurnal fish assemblage, which made up two-thirds of the total number of species at each of the two study sites, is richer and more diversified in food habits than the nocturnal assemblage. The partitioning of space during the diel cycle seems to be more strict and effective than the partitioning of food; all the functional groups (guilds) previously defined were found in both diurnal and nocturnal assemblages, suggesting a good partitioning of habitats and shelters. Conversely, food resources were not uniformly used by reef fishes on a diel rhythm (Harmelin-Vivien, 1979, 1981; Parrish et al., 1985). Motile organisms, such as fish and crustaceans, are usually preyed upon day and night according to two alternative strategies: (1) either different prey species are

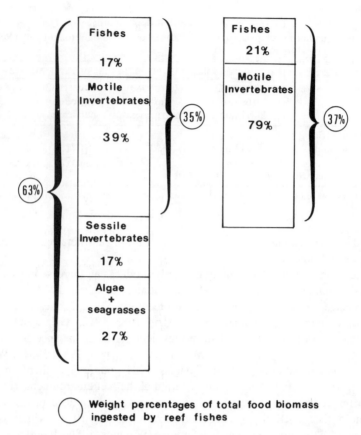

Weight percentages of total food biomass ingested by reef fishes

Figure 2.7. Consumption of the main trophic levels by reef fishes during the day and during the night on Tulear reefs. Low trophic levels are only consumed by day, whereas high trophic levels are used in same quantities by day and by night. (From Harmelin-Vivien, 1981.)

caught by different fish species in the same habitat by day and by night (space overlap, food and time partitioning) or (2) the same prey are caught by different fish species in different habitats (dietary overlap, space and time partitioning). But seagrasses, algae, and sessile invertebrates are only consumed during the day by reef fishes (Figure 2.7). At night they are eaten by motile invertebrates such as gastropods and brachyurans. Thus some food resources remain unused by reef fishes during half of the diel cycle, with invertebrates taking over from them in the food web functioning. Biotic interactions involved in the maintenance of high species richness in coral reefs take place not only within but also between zoological groups. Distinguishing the respective roles of fishes and invertebrates in this interaction is of basic importance for understanding coral reef food webs. For example, algal consumption seems to be mainly due to fishes on Indo-Pacific reefs and to sea urchins on Caribbean reefs, where herbivorous fish diversity is much lower (Ogden and Lobel, 1978).

Biotic Interactions

The principal biotic interactions usually recognized as key processes in community organization are competition for resource utilization, predation, and mutualism. Their importance in coral reef fish assemblages will be briefly discussed in turn.

(1) Interspecific competition for resource utilization (food, space, breeding sites) was long considered the main structuring force in community organization (McArthur, 1965; Pianka, 1966). In this perspective, the high species richness observed in tropical environments was explained either by a decrease in competitive interactions due to a strong specialization of species along one or several of their niche dimensions, or by the presence of overabundant resources, which would make competition pointless. Like all tropical environments, coral reefs are considered to be mainly composed of communities of specialists (Smith and Tyler, 1972, 1973a and b, 1975; Smith, 1975; Gladfelter and Johnson, 1983; Roberts, 1985). An alternative and controversial view makes the reef a world of generalists (Sale, 1977, 1980). To what extent are these hypotheses supported by facts, or not? Space has generally been considered as a limiting factor for reef fishes, and space partitioning has been studied (Smith and Tyler, 1972, 1973a, 1973b; Sale, 1977, 1984; Russell et al., 1974; Ross, 1978; Robertson et al., 1979; and others); it will no longer be discussed here, as I have no personal observations to contribute though I think space might be limiting in some situations. Emphasis will be directed toward food resource utilization and food partitioning by fishes in homologous tropical and temperate environments, i.e., seagrass beds (Harmelin-Vivien, 1983), in order to discuss Sale's view (1984) that "there is no indication that reef fishes divide their resources any more finely than do fish in less diverse, temperate communities" (p. 483). In fact, food resource utilization by fish is more extensive in tropical environments (Harmelin-Vivien, 1983) as observed for birds (McArthur, 1972, and many others). Not only do fish consume a number of food resources unique to the tropics but they also feed on food types present but unused in temperate regions. Dietary categories are indeed much more numerous in the tropics, but for all that are fishes more specialized? Among coral reef fishes, the true specialists are scarce, if we agree with Price's (1984) restricted definition of specialization, which implies a strict dependence of a predator on one particular prey species or even a part of it. Exclusive coral browsers such as certain chaetodontids might be considered as true specialists in Price's sense, but their dependence on corals at a specific or generic level has still to be definitely established (Bouchon-Navarro, 1986). However, "diffuse specialists," defined as species feeding on only one prey category (algae, fishes, brachyurans, sponges, and so on), are more numerous in the tropics than in temperate waters, although they account for no more than 25% of all fish species (Table 2.23). Their percentage of occurrence ranges from 15 to 22% on coral reefs as against 8.5% only in the western Mediterranean. If there are more diffuse specialist fishes in the tropics than elsewhere, generalists with a broad-spectrum diet are also more numerous in tropical environments. The percentages of fish species whose diets include more than 10 prey categories range from 23 to 48% of the fish fauna on coral reefs, and reach only 13% in a temperate region (Table 2.23). Diet spectrum width varies between trophic categories: it is larger in diurnal carnivores, omnivores, and zooplankton feeders, and

Table 2.23. Relative Importance of Trophic "Specialists" and "Generalists" Among Fishes in Tropical and Temperate Environments

Number of Prey Types in Diets	Marseille (b)	Puerto Rico (c)	Tulear (d)	Hawaii (e)	Marshall[a] (f)
1	8.5	18.4	14.8	18.6	21.5
(1–herbivores excluded)	(8.5)	(9.9)	(4.2)	(11.8)	(17.2)
2– 5	42.5	33.0	14.1	28.4	39.9
6–10	36.3	22.2	23.2	24.5	20.6
11–15	6.4	13.7	29.6	25.5	5.6
16–20	6.4	7.6	12.0	7.8	0
21–25	0	1.9	5.6	1.0	0
26–30	0	0	0.7	0	0
Mean no. individuals analyzed/species	22	26	59	15	9 (a)

[a] The low number of generalist fishes found in Marshall Islands may be due to the low number of individuals analyzed for each species.
[b] Bell and Harmelin-Vivien, 1983. [c] Randall, 1967. [d] Harmelin-Vivien, 1979. [e] Hobson, 1974. [f] Hiatt and Strasburg, 1960.

narrower in nocturnal carnivores and sessile invertebrate browsers. However, it is always larger in tropical species within trophic categories (Table 2.24).

The hypothesis of a finer partitioning of food resources in tropical species within a given fish family was tested for the Labridae and Scorpaenidae between species living on the Tulear reef and in the Mediterranean seagrass beds near Marseilles. The general diversity of diet expressed as mean numbers of prey per species and per individual, as well as dietary breadth, did not significantly differ between tropical and temperate species within the labrid family, although slightly higher values were obtained for Tulear species, contrary to the hypothesis tested (Table 2.25). The diet of labrid fishes is more heterogeneous (higher maximum number of prey per species of consumer) on coral reefs, although no relationship is found to exist between habitat diversity and diet diversity in this family. Feeding heterogeneity might simply reflect the larger diversity of resources within coral reefs. Among scorpaenids differences of diet between tropical and temperate species were still smaller and no differences in food types, prey diver-

Table 2.24. Mean Number and Range of Prey Types per Fish Diet Category in Tropical and Temperate Environments[a]

	Marseille (b)	Puerto Rico (c)	Tulear (d)	Hawaii (e)	Marshall (f)
Omnivores	1	10.3(3–20)	16.0(8–25)	14.3(8–17)	5.9(3–11)
Ses. inv. browsers	0	7.3(3–14)	8.1(1–17)	7.8(1–21)	3.6(1–12)
Zooplank. feeders	4.7(2–8)	8.3(2–21)	12.6(7–21)	10.4(5–16)	6.8(2–12)
Carnivores A	7.4(1–16)	10.6(1–19)	13.3(1–26)	8.4(1–20)	6.6(1–15)
Carnivores B	4.1(1–8)	6.5(2–21)	6.7(1–18)	7.5(2–19)	4.1(1–14)

( ) = range.
[a] Herbivores and piscivores considered as "diffuse specialists" are not included.
[b-f] Cf. Table 2.23.

Table 2.25. Comparison of Diet Spectrum in Temperate (Marseille) and Tropical (Tulear) Labrid Fishes

	N	P	p	AH
Marseille	11	11.17 + 3.97 (5–16)	4.25 + 1.26	8.85 + 3.20 (4.06–14.01)
Tulear	16	14.44 + 4.94 (6–24)	3.28 + 1.14	12.72 + 9.24 (3.67–34.12)

N = number of species analyzed; P = mean number of prey types/species; p = mean number of prey types/individual; ( ) = range; AH = diet amplitude ($AH = e^{H'}$, where H' = Shannon index of diversity. Cf. Blondel, 1979).

sity, and diet breadth were found. Results obtained for these two families confirm the general trend observed for broad trophic categories (Table 2.24), but they may not apply to all fish families.

A distinctive characteristic of the trophic structure of fish communities in coral reefs is the striking importance of omnivorous species, as opposed to Stenseth's (1985) view which predicts that omnivory must be rare and optimal food selection the rule in a stable environment.

Studies on food competition among reef fishes are generally based on "average diets." But it has been established that, within a species, diet varies with size, habitat, diel rhythm, and season (Emery, 1973; Vivien and Peyrot-Clausade, 1974; Harmelin-Vivien, 1979; Gladfelter and Johnson, 1983; Wolf, 1985). All these variability factors do not help to unravel the complex biotic interactions taking place between species. To actually demonstrate the existence of competition for food resources, one generally relies on indirect evidence (Thresher, 1983; Bouchon-Navarro, 1986), except in the case of experiments conducted in simple environments, such as those of Werner (1984). The potential interspecific competition is generally measured using different indices: niche breadth, niche similarity, niche overlap, which were long considered as evidences of actual present or past competitive pressures (Levins, 1968; McArthur, 1968; and others), a view presently questioned (cf. Connell, 1974 for a discussion on the "ghost of competition past"). The observed patterns may well reflect independent specific responses to the various characteristics of an heterogeneous environment as much as results from competitive interactions (Connor and Simberloff, 1979; Brown, 1981; Simberloff, 1982; Sale, 1984; Bradley and Bradley, 1985).

To conclude, the increase in both food specialists and broad food generalists in reef fish communities may be considered as an adaptive independent response of species to a complex environment, as well as the result of strong competitive interactions between species. We need well-controlled manipulative experiments to clarify this issue.

(2) Since Paine's paper (1966), predation is considered as a way to reduce interspecific competition by keeping densities of potential competitor populations at low levels. Can we estimate the impact of fish predation on reef fish communities? Some recently conducted experiments (Shulman, 1985b; Victor, 1986) emphasized the role of predators on survival rates of settling juveniles. If true piscivores are not numerous on reefs—4% in Tulear (Harmelin-Vivien, 1979), 6% in Moorea (Galzin, 1985), 1% on some Australian reefs (Williams and Hatcher, 1983)—the total proportion of fish

Table 2.26. Importance of Fish Predation on Different Reef Ichthyofaunas

Prey Types	Puerto Rico (c) N	Puerto Rico (c) %	Tulear (d) N	Tulear (d) %	Hawaii (e) N	Hawaii (e) %	Marshall (f) N	Marshall (f) %
Fish eggs	28	19.7	16	6.9	23	22.6	14	6.6
Fish larvae	5	3.5	3	1.3	–	–	18	8.5
Fish (juv. + adults)	56	39.4	95	40.8	30	29.4	107	50.5
Total no. of fish species analyzed	142		233		102		212	

N = number of fish species; % = percentage in fish species number.
[c-f]Cf. Table 2.23.

species preying at various rates on fish eggs, fish larvae, juvenile and adult reef fishes is high (Table 2.26). Of reef fish species studied by Hiatt and Strasburg (1960), Randall (1967), Hobson (1974), and Harmelin-Vivien (1979), 30–50% preyed on reef fishes and 7–23% fed on benthic or pelagic fish eggs. The important oceanic predation on reef fish larvae by such pelagic species as Tunnids must also be taken into consideration (Fourmanoir, 1969, 1971; Grandperrin, 1975).

In any case, fishes make up an important prey category for reef fishes. On Tulear reefs, fishes did represent 19% of the total weight of prey consumed by the fish community following brachyurans (29%), the major prey item, and algae represented 16% (Harmelin-Vivien, 1979, 1981). Similar trends were observed on Hawaiian reefs where fishes are the third largest major dietary group (10%), after brachyurans (26%) and algae (16%) (Parrish et al., 1985). The predation rate of reef fishes on their own kind must therefore be high, though nonspecialized, as fish predation by fish depends much more on the size and accessibility of the fish prey than on their specific status. Because the density of most of reef fish populations is low, it is tempting to think of a causal relationship between the two phenomena; however, such a relationship is far from being established!

(3) All interactions between species are not of a competitive or predatory nature. Mutualism is one, long underrated type of biotic interaction that greatly contributes to the structuration of communities and is particularly widespread in tropical environments. Its effects on reef fish community structure are often more easily detectable than those of predation. The main mutualistic interactions among reef fishes relate to foraging and cleaning behaviors as well as to protection. Numerous mutualistic interactions in foraging behavior involve fish species with similar diets (Ehrlich and Ehrlich, 1973; Barlow, 1974; Karplus, 1979). Some associations appear to be mutually beneficial as they apparently lead to optimal resource utilization in a heterogeneous environment, or allow the associated species to enter the well-defended territories of sedentary fishes (Lundberg and Lipkin, 1979; Barlow, 1975; Wolf, 1985; Roberts, 1985). Some others seem to provide more asymmetric benefits, as is the case when labrids and lethrinids follow mullid schools, or when *Aulostomus* hunt with groupers (Longley and Hildebrant, 1941; Karplus, 1978; Hobson, 1974). The cleaning behavior of reef fishes, a mutualistic interaction in which the host is cleared of its ectoparasites, while the cleaner is supplied with a regular food source, has often been reported, especially among wrasses (Eibl-Eibesfeld, 1955; Randall, 1958; Limbaugh, 1961; Feder, 1966;

Hobson, 1969; Losey, 1972; Slobodkin and Fishelson, 1974; Kuwamura, 1976; and others). Polyspecific associations with a protective function, such as the diurnal schooling of nocturnal carnivores, are also numerous among reef fishes (Ehrlich and Ehrlich, 1973; Hobson, 1973, 1978).

To sum up, it seems risky to ascribe the maintenance of a high species richness in reef fish communities only to abiotic factors or, conversely, to biotic interactions alone. As noted by Rahel et al. (1984), it seems "simplistic to suggest that assemblages are regulated solely by one process" (p. 583). Indeed, most of the authors, while favoring one process, do not reject the influence of the others (Sale, 1977, 1984; Connell, 1978; Brown, 1981; Bohnsack, 1983). It is likely that all the factors previously discussed, and probably a few others, simultaneously influence community structure, but they do not act on the same level, and with the same intensity, within all communities at the same temporal and spatial scales. At the fish level as well as at the reef level, diversity is the rule and there is no real "reef fish type." All species are not ecologically equivalent, and they differ in their functional roles. Heterogeneity of reef fish communities apparently justifies the often controversial hypotheses put forward to explain the high diversity of coral reef fish communities.

Determinism or Stochasticity? Equilibrium or Nonequilibrium? A Matter of Scale

Two schools of thought confront each other to explain the high species richness of reef fish faunas. They differ essentially in the degree of predictability and of order that they grant to reef assemblages (see Smith, 1978; Sale, 1980, 1984; Bohnsack, 1983; Shulman, 1983; Abrams, 1984; Rahel et al., 1984 for a more comprehensive discussion).

The deterministic school holds the view that high diversity is maintained on reefs through resource partitioning between species, reef fish assemblages being equilibrium communities exhibiting properties of stability, resilience, and persistence (Smith and Tyler, 1972, 1973; Smith, 1975, 1978; Dale, 1978; Robertson and Lassig, 1980; Gladfelter et al., 1980; Ogden and Ebersole, 1981; Grossman, 1982).

The stochastic school considers that the same communities are nonequilibrium, unstable systems, lacking structural persistence. The physiochemical environment is seldom stable enough to allow an equilibrium, and species abundance is determined through independent differential responses to unpredictable environmental changes (Sale, 1974, 1977, 1978, 1980, 1984; Russell et al., 1974; Sale and Dybdahl, 1975; Talbot et al., 1978; Williams, 1980; Sale and Williams, 1982; and others).

Similar viewpoints are expressed, slightly differently, in recent studies of other aquatic or terrestrial communities (e.g., May, 1984; Wiens, 1984 for an overview). Lately, emphasis have been laid on the importance of temporal and spatial scales in ecology and on the pertinence of the scales selected to answer the questions that were formulated (Ogden and Ebersole, 1981; Bohnsack, 1983; May, 1984; Shulman, 1985; Williams, 1986; Blondel, 1986).

Before discussing the scale problem, one must keep in mind that each method generates its own particular error. This inescapable background noise is often ignored or not taken into account. This problem is very serious in visual estimations of reef fish populations, for instance, where only the visible part of the community can be enumerated (see Thresher, 1983; Harmelin-Vivien et al., 1985; and Thresher and Gunn, 1986 for a review of these problems). In all cases, an intrinsic error is made, which depends on the probability of a species to be seen by the diver, a probability modified by transect width, home range size, fish behavior, and fish–diver interactions. The influence of sampling on data variability and representativity has been often discussed (Frontier, 1983) and still remains a serious problem. Data variance is not only due to the processes studied abut also to the methods and species observed. Their respective contributions are often difficult to evaluate, and this raises problems for those concerned with the dynamics of fluctuation.

Numerous authors (Sousa, 1979; Wiens, 1984; May, 1984) have noticed that systematic patterns, or evidence of global equilibrium, in the specific composition of communities generally appear when systems are examined on a sufficiently large spatial or temporal scale. No such patterns are apparent at smaller scales where nonequilibrium trends generally predominate.

Spatial Scale

The application of the fractal theory to reef ecology suggests the existence of three levels of perception. Within each level the processes generating the reef topography are self-similar: the individual structure of coral colonies (< 10 cm), the coral colonies (20–200 cm), and the geomorphological structures (> 5 m) (Bradbury et al., 1984). However, we have seen that the highly positive correlation between coral and fish species richness observed on a biogeographic scale may disappear at a reef scale, or may be found in one geomorphological zone and not in the other. Galzin (1985, 1987a) and Galzin and Legendre (1988) conducted a study to define the spatial scale at which a community structure could be identified according to the null hypothesis that fish species distribution could be random. They concluded that reef fishes were distributed in different communities on a finer spatial scale than the geomorphological one usually referred to. Nevertheless a different problem of scale arose as the number of communities identified depended on the alpha level of significance used for mathematical analysis. On the same transect, four groups were partitioned at the alpha = 5% level of significance and seven groups were found when alpha reaches 30% (Galzin and Legendre, 1988).

Temporal Scale

Bohnsack (1983) showed that the time interval between sampling might have a great influence on the estimation of species turnover rate and thereafter on conclusions drawn on the reef fish community dynamics. The authors studying recruitment patterns (Sale, Doherty et al., 1984; Sale, 1984; Thresher, 1983; Williams, 1983, 1986) emphasize the striking temporal variations in reef fish recruitment rates, whatever may be the time scale considered: week, month, or year. Conversely, Kock (1982), Molina (1983), and Galzin (1985), while studying the annual variations of well-established fish communities, found that they remain stable on a yearly cycle, although they evidenced seasonal fluctuations. Galzin (1985) also demonstrated that the rates of variation in community structure differed at different time scales. The higher rates were observed between diurnal and nocturnal communities as a result of the differences in activity rhythm and behavior of the various species. Rates of variation were much lower for monthly and annual cycles. But the coefficient of variation for species abundance was higher for the annual cycle (0.13) than for the monthly cycle (0.06), although always low. These annual variations are simultaneously induced by reproduction, recruitment, and predation patterns.

Community Definition

The intepretation of results may also depend on how the assemblage is defined (Rahel et al., 1984; Walsh, 1985; Cameron and Endean, 1985). The studies conducted by Grossman et al. (1982) and Rahel et al. (1984) on this matter are highly suggestive. Grossman et al. (1982) defined the fish assemblage of Otter Creek as the 10 most abundant species within each season. They concluded that this assemblage was stochastic, and not in equilibrium, and that it was mainly regulated by abiotic processes. Reconsidering the same data but defining the fish assemblage as all the species collected at all seasons, Rahel et al. (1984) found that the fish assemblage structure of Otter Creek was remarkably persistent! They concluded their paper by pointing out that "whether a particular assemblage appears to be regulated by stochastic or deterministic processes may well depend on how one defines that assemblage" (p. 588). They specified that these processes may change with species and the age of individuals. Abiotic factors and stochastic processes may well be most prominent for fish larvae and small juveniles, whereas older age classes may be regulated more by biotic interactions and deterministic processes.

To conclude, it appears that structural patterns and regulation processes of coral reef fish assemblages are not only more varied and complex than expected, but that the views we can hold on them depend on the definition of the assemblage itself and largely on the spatial and temporal scales chosen to test the hypotheses. More than just one cause or one mechanism, one should consider a group of strikingly different causes in which the hierarchical relationships may vary. Complex structural patterns require complex explanatory hypotheses and Brown's (1981) consideration on community structure is always valid, i.e., "the relationship between species diversity, functional organization and stability of community remains a challenging problem" (p. 880).

Acknowledgments. Greatest thanks are expressed to Dr. M. Pichon, Australian Institute of Marine Science, Townsville, for a critical review of the manuscript.

References

Abele, L.G. 1984. Biogeography, colonization and experimental community structure of coral-associated crustaceans. In: Strong, D.R. Jr., Simberloff, D., Abele, L.G., and Thistle, A.B. (eds.), Ecological Communities: Conceptual Issues and the Evidence, Princeton Univ. Press, Princeton, NJ, pp. 123–137

Abrams, P.A. 1984. Recruitment, lotteries and coexistence in coral reef fish. Am. Nat. 123(1), 44–55

Allen, G.R. 1975. Damselfishes of the South Seas. T.F.H. Publ., Neptune City, NJ

Anderson, G.R.V., Ehrlich, A.H., Ehrlich, P.R., Roughgarden, J.D., Russell, B.C., and Talbot, F.H. 1981. The community structure of coral reef fishes. Am. Nat. 117(4), 476–495

Barlow, G.W. 1974. Extraspecific imposition of social grouping among surgeonfishes (Pisces: Acanthuridae). J. Zool. London 174, 333–340

Barlow, G.W. 1975. On the sociology of four Puerto-Rican parrotfishes (Scaridae). Mar. Biol. 33(4), 281–294

Bell, J.D., and Galzin, R. 1984. The influence of live coral cover on coral reef fish communities. Mar. Ecol. Prog. Ser. 15(3), 265–274

Bell, J.D., and Harmelin-Vivien, M.L. 1983. Fish fauna of french Mediterranean *Posidonia oceanica* seagrass meadows. II. Feeding habits. Téthys 11(1), 1–14

Blondel, J. 1979. Biogéographie et Ecologie, Masson, Paris

Blondel, J. 1986. Biogéographie évolutive, Masson, Paris

Bohnsack, J.A. 1983. Species turnover and the order versus chaos controversy concerning reef fish community structure. Coral Reefs 1(4), 223–228

Bouchon, C. 1981. Quantitative study of the scleractinian coral communities of a fringing reef of Reunion Island (Indian Ocean). Mar. Ecol. Prog. Ser. 4(3), 273–288

Bouchon, C. 1985. Quantitative study of Scleractinian coral communities of Tiahura reef (Moorea Island, French Polynesia). Proc. 5th Int. Coral Reef Symp. 6, 279–284

Bouchon, C., and Bouchon-Navarro, Y. 1981a. Etude d'environnement de la Baie de la Possession. Rapport, Centre Univ. La Réunion

Bouchon, C., and Bouchon-Navarro, Y. 1981b. Etude d'environnement du lagon du récif de l'Hermitage (Lieu-dit: GO PAYET). Rapport, Centre Univ. La Réunion

Bouchon-Navarro, Y. 1981. Quantitative distribution of the Chaetodontidae on a reef of Moorea Island (French Polynesia). J. Exp. Mar. Biol. Ecol. 55, 145–157

Bouchon-Navarro, Y. 1986. Partitioning of food and space resources by Chaetodontid fishes on coral reefs. J. Exp. Mar. Biol. Ecol. 103, 21–40

Bouchon-Navarro, Y., Bouchon, C., and Harmelin-Vivien, M.L. 1985. Impact of coral degradation on a chaetodontid fish assemblage (Moorea, French Polynesia). Proc. 5th Int. Coral Reef Symp. 5, 427–432

Bourlière, F. 1983. Animal species diversity in tropical forests. In: Golley, F.B., Lieth, H., and Werger, M.J.A. (eds.), Tropical Rain Forest Ecosystems, Elsevier, Amsterdam, pp. 77–91

Bradbury, R.H., Reichelt, R.E., and Green, D.G. 1984. Fractals in ecology: Methods and interpretation. Mar. Ecol. Prog. Ser. 14, 295–296

Bradley, R.A., and Bradley, D.W. 1985. Do non-random patterns of species in niche space imply competition? Oikos 45(3), 443–446

Brown, J.T. 1981. Two decades of homage to Santa Rosalia: Toward a general theory of diversity. Am. Zool. 21, 877–888

Cameron, A.M., and Endean, R. 1985. Do long-lived species structure coral reef ecosystems? Proc. 5th Int. Coral Reef Symp. 6, 211–215

Charles-Dominique, P. 1975. Nocturnality and diurnality. An ecological interpretation of these two modes of life by an analysis of the higher vertebrate fauna in tropical forest ecosystems. In: Luckett, W.P., and Szalay, F.S. (eds.), Phylogeny of Primates, Plenum Press, New York, pp. 69–88

Cody, M.L., and Diamond, J.M. 1975. Ecology and Evolution of Communities. Belknap Press, Cambridge, MA

Comins, H.N., and Noble, I.R. 1985. Dispersal, variability, and transient niches: species coexistence in a uniformly variable environment. Am. Nat. 126(5), 706–723

Connell, J.H. 1974. Ecology: Field experiments in marine ecology. In: Mariscal, R. (ed.), Experimental Marine Biology, Academic Press, New York, pp. 21–54

Connell, J.H. 1978. Diversity in tropical rain forests and coral reefs. Science 199, 1302–1309

Connell, J.H. 1983. Disturbance and patch dynamics of reef-corals. In: Baker, J.T., Carter, R.M., Sammarco, P.W., and Stark, K.P. (eds.), Proc. Great Barrier Reef Conference, 1983, pp. 179–189

Connell, J.H., and Orias, E. 1964. The ecological regulation of species diversity. Am. Nat. 98, 399–414

Connor, E.F., and Simberloff, D. 1979. The assembly of species communities: chance or competition? Ecology 60(6), 1132–1140

Dale, G. 1978. Money-in-the-bank: A model for coral reef fish coexistence. Env. Biol. Fish 3(1), 103–108

Dauget, J.M. 1986. Application des méthodes architecturales aux coraux, quelques traits communs aux formes vivantes fixées. Thèse Doc., Univ. Sciences et Techniques du Languedoc

Downing, N. 1985. Coral reef communities in an extreme environment: The northwestern arabian gulf. Proc. 5th Int. Coral Reef Symp. 6, 343–348

Ehrlich, P.R., and Ehrlich, A.H. 1973. Coevolution: Heterotypic schooling in caribbean reef fishes. Am. Nat. 107(953), 157–160

Eibl-Eibesfeldt, I. 1955. Über Symbiosen, Parasitismus und andere besondere zwischenartliche Beziehungen tropischer Meeresfische. Zeit. Tierpsychol. 12(2), 203–219

Elton, C.S. 1958. The Ecology of Invasion by Animals and Plants, Methuen, London

Emery, A.R. 1973. Comparative ecology and functional osteology of fourteen species of damselfish (Pisces: Pomacentridae) at Alligator Reef, Florida Keys. Bull. Mar. Sci. 23(3), 649–770

Feder, H.M. 1966. Cleaning symbiosis in the marine environment. In: Henry, S.M. (ed.), Symbiosis, Vol. 1, Academic Press, New York, pp. 327–380

Fourmanoir, P. 1969. Contenus stomacaux d'Alepisaurus (poissons) dans le sud-ouest du Pacifique. Cah. O.R.S.T.O.M., Sér. Océanogr. 7(4), 51–60

Fourmanoir, P. 1971. Liste des espèces de poissons contenus dans les estomacs de thons jaunes, Thunnus albacares (Bonnaterre, 1788) et de thons blancs, Thunnus alalunga (Bonnaterre, 1788). Cah. O.R.S.T.O.M., Sér. Océanogr. 9(2), 109–118

Fourmanoir, P., and Laboute, P. 1976. Poissons de Nouvelle-Calédonie et des Nouvelles Hébrides. Les Editions du Pacifique, Papeete

Frontier, S. (ed.). 1983. Stratégies d'échantillonnage en Ecologie. Masson, Paris

Gabrié, C., and Salvat, B. 1985. General features of French Polynesian islands and their coral reefs. Proc. 5th Int. Coral Reef Symp. 1, 1–15

Galzin, R. 1977. Richesse et productivité des écosystèmes lagunaires et récifaux. Applications à l'étude dynamique d'une population de Pomacentrus nigricans du lagon de Moorea (Polynésie Française). Thèse 3ème cycle, Univ. Sciences et Techniques du Languedoc

Galzin, R. 1979. La faune ichtyologique d'un récif corallien de Moorea, Polynésie Française: Échantillonnage et premiers résultats. Terre Vie, Rev. Ecol. 33, 623–643

Galzin, R. 1985. Ecologie des poissons récifaux de Polynésie Française: Variations spatiotemporelles des peuplements; dynamique de populations de trois espèces dominantes des lagons nord de Moorea; évaluation de la production ichtyologique d'un secteur récifolagonaire. Thèse Doc. ès-Sci., Univ. Sciences et Techniques du Languedoc

Galzin, R. 1987a. Structure of fish communities of French Polynesian coral reefs. I. Spatial scales. Mar. Ecol. Prog. Ser. 41, 129–136

Galzin, R. 1987b. Structure of fish communities of French Polynesian coral reefs. II. Temporal scales. Mar. Ecol. Prog. Ser. 41, 137–145

Galzin, R., and Legendre, P. 1988. The fish communities of a coral reef transect. Pac. Sci. 41(1–4), 158–165

Gladfelter, W.B., and Gladfelter, E.H. 1978. Fish community structure as a function of habitat structure on West Indian patch reefs. Rev. Biol. Trop. 26 (suppl. 1), 65–84

Gladfelter, W.B., and Johnson, W.S. 1983. Feeding niche separation in a guild of tropical reef fishes (Holocentridae). Ecology 64(3), 552–563

Gladfelter, W.B., Ogden, J.C., and Gladfelter, E.H. 1980. Similarity and diversity among patch reef fish communities: A comparison between tropical western Atlantic (Virgin Islands) and tropical central Pacific (Marshall Islands) patch reefs. Ecology 61(5), 1156–1168

Goldman, B. 1973. Ecology of One Tree reef fishes. Ph.D. thesis, Univ. Sydney

Goldman, B., and Talbot, F.H. 1976. Aspects of the ecology of coral reef fishes. In: Jones, O.A., and Endean, R. (eds.), Biology and Geology of Coral Reefs, Vol. 3, Biology 2, Academic Press, New York, pp. 125–154

Gosline, W.A. 1971. Functional Morphology and Classification of Teleostean Fishes, Univ. Hawaii Press, Honolulu

Grandperrin, R. 1975. Structures trophiques aboutissant aux thons de longue ligne dans le Pacifique sud-ouest tropical. Mém. O.R.S.T.O.M., Paris

Grossman, G.D. 1982. Dynamics and organization of a rocky intertidal fish assemblage: the persistence and resilience of taxocene structure. Am. Nat. 119, 611–637

Grossman, G.D., Moyle, P.B., and Whitaker, J.O. Jr. 1982. Stochasticity in structural and functional characteristics of an indiana stream fish assemblage: A test of community theory. Am. Nat. 120, 423–454

Harmelin-Vivien, M.L. 1976. Ichtyofaune de quelques récifs coralliens des iles Maurice et La Réunion (Archipel des Mascareignes, Océan Indien). Mauritius Inst. Bull. 8(2), 69–104

Harmelin-Vivien, M.L. 1979. Ichtyofaune des récifs coralliens de Tuléar (Madagascar): Ecologie et relations trophiques. Thèse Doc. ès-Sci., Univ. Aix-Marseille II

Harmelin-Vivien, M.L. 1981. Trophic relationships of reef fishes in Tuléar (Madagascar). Oceanologica Acta 4(3), 365–374

Harmelin-Vivien, M.L. 1983. Etude comparative de l'ichtyofaune des herbiers de Phanérogames marines en milieu tropical et tempéré. Rev. Ecol. (Terre Vie) 38(2), 179–210

Harmelin-Vivien, M.L., and Bouchon-Navarro, Y. 1983. Feeding diets and significance of coral feeding among Chaetodontid fishes in Moorea (French Polynesia). Coral Reefs 2, 119–127

Harmelin-Vivien, M.L., Harmelin, J.G., Chauvet, C., Duval, C., Galzin, R., Lejeune, P., Barnabe, G., Blanc, F., Chevalier, R., Duclerc, J., and Lasserre, G. 1985. Evaluation visuelle des peuplements et populations de poissons: Méthodes et problèmes. Rev. Ecol. (Terre Vie) 40, 467–539

Harrison, P.L., Babcock, R.C., Bull, G.D., Oliver, J.K., Wallace, C.C., and Willis, B.L. 1984. Mass spawning in tropical reef corals. Science 223, 1186–1189

Hiatt, R.W., and Strasburg, D.W. 1960. Ecological relationships of the fish fauna on coral reefs of the Marshall Islands. Ecol. Monogr. 30, 65–127

Hobson, E.S. 1969. Comments on certain recent generalizations regarding cleaning symbiosis in fishes. Pac. Sci. 23(1), 35–39

Hobson, E.S. 1973. Diel feeding migrations in tropical reef fishes. Helgoländer wiss. Meeresunters. 24(1–4), 361–370

Hobson, E.S. 1974. Feeding relationships of Teleostean fishes on coral reefs in Kona, Hawaii. Fish. Bull. 72(4), 915–1031

Hobson, E.S. 1978. Aggregating as a defense against predators in aquatic and terrestrial environments. In: Reese, E.S., and Lighter, F.J. (eds.), Contrasts in Behavior, Wiley Interscience, New York, pp. 219–234

Huston, M. 1979. A general hypothesis of species diversity. Am. Nat. 113, 81–101

Hutchinson, G.E. 1959. Homage to Santa Rosalia or why are there so many kinds of animals? Am. Nat. 93, 145–159

Jones, R.S. 1968. Ecological relationships in Hawaiian and Johnston Island Acanthuridae (surgeonfishes). Micronesica 4(2), 309–361

Karplus, I. 1979. The tactile communication between Cryptocentrus steinitzi (Pisces, Gobiidae) and Alpheus purpurilenticularis (Crustacea, Alpheidae). Z. Tierpsychol. 49(2), 173–196

Kock, R.L. 1982. Patterns of abundance variation in reef fishes near an artificial reef in Guam. Env. Biol. Fish. 7(2), 121–136

Kuwamura, T. 1976. Different responses of inshore fishes to the cleaning wrasse, Labroides dimidiatus, as observed in Sirahama. Publ. Seto Mar. Biol. Lab. 23(1–2), 119–144

Lawton, J.H. 1982. Vacant niches and unsaturated communities: Comparison of bracken herbivores at sites on two continents. J. Anim. Ecol. 51, 573–595

Lawton, J.H. 1984. Non-competitive populations, non-convergent communities, and vacant niches: The herbivores of bracken. In: Strong, D.R. Jr., Simberloff, D., Abele, L.G., and Thistle, A.B. (eds.), Ecological Communities: Conceptual Issues and the Evidence, Princeton Univ. Press, Princeton, NJ, pp. 67–99

Levins, R. 1968. Evolution in Changing Environments, Princeton Univ. Press, Princeton, NJ

Limbaugh, C. 1961. Cleaning symbiosis. Sci. Am. 205(2), 42–49

Longley, W.H., and Hildebrand, S.F. 1941. Systematic catalogue of the fishes of Tortugas, Florida with observations on color, habits and local distribution. Pap. Tortugas Lab. 34, 1–331

Losey, G.S. 1972. The ecological importance of cleaning symbiosis. Copeia 1972(4), 820–833

Loya, Y. 1985. Seasonal changes in growth rate of a Red Sea coral population. Proc. 5th Int. Coral Reef Symp. 6, 187–191

Luckhurst, B.E., and Luckhurst, K. 1978. Analysis of the influence of substrate variables on coral reef fish communities. Mar. Biol. 49(4), 317–323

Lundberg, B., and Lipkin, Y. 1979. Natural food of the herbivorous rabbit fish (*Siganus* sp.) in the northern Red Sea. Botanica Marina 22, 173–183

Mather, P., and Bennett, I. (eds.). 1984. A Coral Reef Handbook. A Guide to the Fauna, Flora and Geology of Heron Island and Adjacent Reefs and Cays. The Australian Coral Reef Society, Handbook Series, Vol. 1

Maugé, A.L. 1967. Contribution préliminaire à l'inventaire ichtyologique de la région de Tuléar. Rec. Trav. Sta. mar. Endoume, fasc. hors sér., suppl. 7, 101–132

May, R.M. 1984. An overview: Real and apparent patterns in community structure. In: Strong, D.R. Jr., Simberloff, D., Abele, L.G., and Thistle, A.B. (eds.), Ecological Communities: Conceptual Issues and the Evidence, Princeton Univ. Press, Princeton, NJ, pp. 3–16

McArthur, R.H. 1965. Patterns of species diversity. Biol. Rev. (Camb. Philos. Soc.) 40, 510–533

McArthur, R.H. 1968. The theory of the niche. In: Lewontin, R.C. (ed.), Population Biology and Evolution, Syracuse Univ. Press, Syracuse, NY, pp. 159–176

McArthur, R.H. 1972. Geographical Ecology: Patterns in the Distribution of Species. Harper and Row, New York

McArthur, R.H., and Wilson, E.O. 1967. The Theory of Island Biogeography. Princeton Univ. Press, Princeton, NJ

Molina, M.M. 1983. Seasonal and annual variation of coral-reef fishes on the upper reef slope at Guam. M. Sc. Biology, Univ. Guam

Molles, M.C. Jr. 1978. Fish species diversity on model and natural reef patches: Experimental insular biogeography. Ecol. Monogr. 48, 289–305

Morrissey, J. 1985. Primary productivity of coral reef benthic macroalgae. Proc. 5th Int. Coral Reef Symp. 5, 77–82

Ogden, J.C., and Ebersole, J.P. 1981. Scale and community structure of coral reef fishes: a long-term study of a large artificial reef. Mar. Ecol. Prog. Ser. 4(1), 97–103

Ogden, J.C., and Lobel, P.S. 1978. The role of herbivorous fishes and urchins in coral reef communities. Env. Biol. Fish 3(1), 49–63

Paine, R.T. 1966. Food web complexity and species diversity. Am. Nat. 100, 65–75

Parrish, J.D., Callahan, M.W., and Norris, J.E. 1985. Fish trophic relationships that structure reef communities. Proc. 5th Int. Coral Reef Symp. 4, 73–78

Payri, C.E. 1982. Les macrophytes du Lagon de Tiahura (Ile de Moorea, Polynésie Française): Inventaire floristique, Répartition, Biomasses, Variations saisonnières, Dynamique des populations de *Turbinaria ornata* (Phéophycées, Fucales). Thèse 3ème cycle, Univ. Sciences et Techniques du Languedoc

Pearson, D.L. 1977. A pantropical comparison of bird community structure on six lowland forest sites. The Condor 79, 232–244

Peyrot-Clausade, M. 1976. Polychètes de la cryptofaune du récif de Tiahura (Moorea). Cah. Pac. 19, 325–337

Peyrot-Clausade, M. 1977a. Faune cavitaire mobile des platiers coralliens de la région de Tuléar (Madagascar). Thèse Doc. ès-Sci., Univ. Aix-Marseille II

Peyrot-Clausade, M. 1977b. Décapodes Brachyoures et Anomoures (à l'exclusion des Paguridae) de la cryptofaune du récif de Tiahura (Moorea). Cah. Pac. 20, 211–222

Pianka, E.R. 1966. Latitudinal gradients in species diversity: A review of concepts. Am. Nat. 100, 33–46

Pianka, E.R. 1974. Evolutionary Ecology, Harper and Row, New York

Pichon, M. 1978. Recherches sur les peuplements à dominance d'Anthozoaires dans les récifs coralliens de Tuléar (Madagascar). Atoll. Res. Bull. 222, 1–447

Pichon, M. 1985. Scleractinians. In: Richard, G. (ed.), Fauna and Flora: A First Compendium of French Polynesian Sea-Dwellers. Proc. 5th Int. Coral Reef Symp. 1, 399–403

Price, P.W. 1980. Evolutionary Biology of Parasites, Princeton Univ. Press, Princeton, NJ

Price, P.W. 1984. Communities of specialists: Vacant niches in ecological and evolutionary time. In: Strong, D.R. Jr., Simberloff, D., Abele, L.G., and Thistle, A.B. (eds.), Ecological Communities: Conceptual Issues and the Evidence, Princeton Univ. Press, Princeton, NJ, pp. 510–523

Rahel, F.J., Lyons, J.D., and Cochran, P.A. 1984. Stochastic or deterministic regulation of assemblage structure? It may depend on how the assemblage is defined. Am. Nat. 124, 583–589

Randall, J.E. 1958. A review of the Labrid fish genus Labroides, with description of two new species and notes on ecology. Pac. Sci. 12(4), 327–347

Randall, J.E. 1967. Food habits of reef fishes of the West Indies. Stud. Trop. Oceanogr. (Miami) 5, 665–847

Randall, J.E. 1985. Fishes. In: Richard, G. (ed.)., Fauna and Flora: A First Compendium of French Polynesia Sea Dwellers. Proc. 5th Int. Coral Reef Symp. 1, 462–481

Ricklefs, R.E. 1979. Ecology, 2nd ed., Nelson, Sunbury-on-Thames

Risk, M.J. 1972. Fish diversity on a coral reef in the Virgin Islands. Atoll Res. Bull. 153, 1–7

Roberts, C.M. 1985. Resource sharing in territorial herbivorous reef fishes. Proc. 5th Int. Coral Reef Symp. 4, 17–22

Robertson, D.R., and Lassig, B. 1980. Spatial distribution patterns and coexistence of a group of territorial damselfishes from the Great Barrier Reef Bull. Mar. Sci. 30, 187–203

Robertson, D.R., Polunin, N.V.C., and Leighton, K. 1979. The behavioral ecology of three Indian Ocean surgeonfishes (Acanthurus lineatus, A. leucosternon and Zebrasoma scopas): Their feeding strategies, and social and mating systems. Env. Biol. Fish. 4(2), 125–170

Robin, B., Petron, C., and Rives, C. 1980. Les coraux: Nouvelle-Calédonie, Tahiti, Réunion, Antilles. Les Editions du Pacifique, Papeete

Rohde, K. 1978. Latitudinal gradients in species diversity and their causes. I. A review of the hypotheses explaining the gradients. Zbl. Biol. 97, 393–403

Ross, R.M. 1978. Reproductive behavior of the Anemonefish Amphiprion melanopus on Guam. Copeia 1978(1), 103–107

Russell, B.C., Talbot, F.H., and Domm, S. 1974. Patterns of colonisation of artificial reefs by coral reef fishes. Proc. 2nd Int. Coral Reef Symp. 1, 207–215

Sale, P.F. 1972. Influence of corals in the dispersion of the Pomacentrid fish Dascyllus aruanus. Ecology 53(4), 741–744

Sale, P.F. 1977. Maintenance of high diversity in coral reef fish communities. Am. Nat. 111, 337–359

Sale, P.F. 1978. Coexistence of coral reef fishes: A lottery for living space. Env. Biol. Fish. 3(1), 85–102

Sale, P.F. 1980. The ecology of fishes on coral reefs. Oceanogr. Mar. Biol. Annu. Rev. 18, 367–421

Sale, P.F. 1984. The structure of communities of fish on coral reefs and the merit of a hypothesis-testing. Manipulative approach to ecology. In: Strong, D.R. Jr., Simberloff, D., Abele, L.G., and Thistle, A.B. (eds.), Ecological Communities: Conceptual Issues and the Evidence, Princeton Univ. Press, Princeton, NJ, pp. 478–490

Sale, P.F., Doherty, P.J., Eckert, G.J., Douglas, W.A., and Ferrell, D.J. 1984. Large-scale spatial and temporal variation in recruitment to fish populations on coral reefs. Oecologia 64(2), 191–198

Sale, P.F., and Douglas, W.A. 1984. Temporal variability in the community structure of fish on coral patch reefs and the relation of community structure to reef structure. Ecology 65, 409–422

Sale, P.F., Douglas, W.A., and Doherty, P.J. 1984. Choice of microhabitats by coral reef fishes at settlement. Coral Reefs 3(2), 91–100

Sale, P.F., and Dybdahl, R. 1975. Determinants of community structure for coral reef fishes in an experimental habitat. Ecology 56(6), 1343–1355

Sale, P.F., and Williams, D.Mc.B. 1982. Community structure of coral reef fishes: Are the patterns more than those expected by chance? Am. Nat. 120, 121–127

Schlesinger, Y., and Loya, Y. 1985. Coral community reproductive patterns: Red Sea versus the Great Barrier Reef. Science 228, 1333–1335

Shulman, M.J. 1983. Species richness and community predictability in coral reef fish faunas. Ecology 64(5), 1308–1311

Shulman, M.J. 1985a. Variability on recruitment of coral reef fishes. J. Exp. Mar. Biol. Ecol. 89, 205–219

Shulman, M.J. 1985b. Recruitment of coral reef fishes; effects of distribution of predators and shelter. Ecology 66(3), 1056–1066

Simberloff, D. 1982. The status of competition theory in ecology. Ann. Zool. Fenn. 19, 241–253

Slobodkin, L.B., and Fishelson, L. 1974. The effect of the cleaner fish *Labroides dimidiatus* on the point diversity of fishes on the reef front at Eilat, Red Sea. Am. Nat. 108, 369–376

Smith, C.L. 1975. Analysis of a coral reef fish community: Size and relative abundance. Hydro-Lab. J. 3(1), 31–38

Smith, C.L. 1978. Coral reef fish communities: A compromise view. Env. Biol. Fish. 3(1), 109–128

Smith, C.L., and Tyler, J.C. 1972. Space resource sharing in a coral reef fish community. Bull. Nat. His. Mus. Los Angeles, Sci. Bull. 14, 125–170

Smith, C.L., and Tyler, J.C. 1973a. Population ecology of a Bahamian suprabenthic shore fish assemblage. Am. Mus. Novit. 2528, 1–38

Smith, C.L., and Tyler, J.C. 1973b. Direct observations of resource sharing in coral reef fish. Helgoländer wiss. Meersunters. 24(1–4), 264–275

Smith, C.L., and Tyler, J.C. 1975. Succession and stability in fish communities of dome-shaped patch reefs in the West Indies. Am. Mus. Novit. 2572, 1–18

Sournia, A. 1972 (1973). Productivité primaire dans le canal de Mozambique. J. Mar. Biol. Assoc. India 14(1), 139–147

Sournia, A., and Ricard, M. 1976. Phytoplankton and its contribution to primary productivity in two coral reef areas of French Polynesia. J. Exp. Mar. Biol. Ecol. 21, 129–140

Sousa, W.P. 1979. Disturbance in marine intertidal boulder fields: The non-equilibrium maintenance of species diversity. Ecology 60, 1225–1239

Stenseth, N.C. 1985. The structure of food webs predicted from optimal food selection models: An alternative to Pimm's stability hypothesis. Oikos 44(2), 361–364

Strong, D.R., Simberloff, D., Abele, L.G., and Thistle, A.B. 1984. Ecological communities: Conceptual issues and the evidence. Princeton Univ. Press, Princeton, NJ

Talbot, F.H. 1965. A description of the coral structure of Tutia reefs (Tanganika territory, East Africa) and its fish fauna. Proc. Zool. Soc. London 145(4), 431–470

Talbot, F.H., and Goldman, B. 1972. A preliminary report on the diversity and feeding relationships of the reef fishes at One Tree Island, Great Barrier Reef system. In: Proc. Symp. on Corals and Coral Reefs, 1969, J. Mar. Biol. Assoc. India, 425–444

Talbot, F.H., Russel, B.C., and Anderson, G.R.V. 1978. Coral reef-fish communities: Unstable, high diversity systems? Ecol. Monogr. 48, 425–440

Talbot, F., and Steene, R. (eds.). 1984. Reader's Digest Book of the Great Barrier Reef. Reader's Digest Publ., Sydney, Australia

Thresher, R.E. 1983. Habitat effects on reproductive success in the coral reef fish, *Acanthochromis polyacanthus* (Pomacentridae). Ecology 64(5), 1184–1199

Thresher, R.E., and Gunn, J.S. 1986. Comparative analysis of visual census techniques for highly mobile, reef-associated piscivores (Carangidae). Envir. Biol. Fish. 17(2), 93–116

Victor, B. 1986. Larval settlement and juvenile mortality in a recruitment-limited coral reef fish population. Ecol. Monogr. 56(2), 145–160

Vivien, M.L. 1973. Contribution à la connaissance de l'éthologie alimentaire de l'ichtyofaune du platier interne des récifs coralliens de Tuléar (Madagascar). Téthys, suppl. 5, 221–308

Vivien, M.L., and Peyrot-Clausade, M. 1974. Comparative study of the feeding behaviour of three coral reef fishes (Holocentridae), with special reference to the Polychaeta of the reef cryptofauna, as prey. Proc. 2nd Int. Coral Reef Symp. 1, 179–192

Walsh, W.J. 1985. Reef fish community dynamics on small artificial reefs: The influence of isolation, habitat structure and biogeography. Bull. Mar. Sci. 36(2), 357–376

Werner, E.E. 1984. The mechanisms of species interactions and community organization in fish. In: Strong, D.R. Jr., Simberloff, D., Abele, L.G., and Thistle, A.B. (eds.), Ecological Communities: Conceptual Issues and the Evidence, Princeton Univ. Press, Princeton, NJ, pp. 360–382

Wiens, J.A. 1984. On understanding a non-equilibrium world: Myth and reality in community patterns and processes. In: Strong, D.R., Simberloff, D., Abele, L.G., and Thistle, A.B. (eds.), Ecological Communities: Conceptual Issues and the Evidence, Princeton Univ. Press, Princeton, NJ, pp. 439–457

Williams, D.Mc.B. 1980. The dynamics of the Pomacentrid community on small patch reefs in One Tree lagoon (Great Barrier Reef). Bull. Mar. Sci. 30, 159–170

Williams, D.McB. 1983. Daily, monthly and yearly variability in recruitment of a guild of coral reef fishes. Mar. Ecol. Prog. Ser. 10(3), 231–237

Williams, D.McB. 1986. Temporal variation in the structure of reef slope fish communities (Central Great Barrier Reef): Short-term effects of *Acanthaster planci* infestation. Mar. Ecol. Prog. Ser. 28, 157–164

Williams, D.Mc.B., and Hatcher, A.I. 1983. Structure of fish communities on outer slopes of inshore, mid-shelf and outer shelf reefs of the Great Barrier Reef. Mar. Ecol. Prog. Ser. 10(3), 239–250

Whittaker, R.H. 1977. Evolution of species diversity in land communities. In: Hecht, M.K., Steere, W.C., and Wallace, B. (eds.), Evol. Biol. 10, Plenum Press, New York, pp. 1–67

Wolf, N.G. 1985. Food selection and resources partitioning by herbivorous fishes in mixed-species groups. Proc. 5th Int. Coral Reef Symp. 4, 23–28

3. Tropical Herpetofaunal Communities: Patterns of Community Structure in Neotropical Rainforests

William E. Duellman

Tropical rainforests or "jungles" are commonly thought of as harboring a vast array of man-eating crocodiles, gaudy lizards, noisy frogs, and huge snakes which either "crush" their victims or dispense with them quickly by injection of lethal venoms. This view of the rainforests, so commonly portrayed in Hollywood productions, contains less truth than it does fiction. In contrast, the scientific literature dealing with herpeto-faunal communities of tropical rainforests is rich in factual information, the interpretation of which frequently is more fictional than true.

In a review of resource partitioning in ecological communities, Schoener (1974) identified three important ways in which similar species partition resources: (1) habitat, (2) diet, and (3) time; he concluded that the latter was the least important. Schoener (1974) noted that there are substantial differences in methods of resource partitioning among terrestrial ectotherms and endotherms. According to Pough (1980), these differences derive in part from the great difference in modal body size between ectothermic and endothermic tetrapods. Because of basic differences in physiology and life history, ectothermic vertebrates are likely to respond differently from endotherms to habitat dimensions, and temporal partitioning may be much more important in ectotherms.

The following is a synthesis of the factual information about herpetofaunal communities in neotropical rainforests and an interpretation of these data in light of existing theory. Principally, I am attempting to define patterns of resource utilization in the communities and to determine if these patterns are replicated in rainforests throughout the neotropics.

I emphasize that I hold no particular "school" of community ecology sacred and that the factors shaping communities of ectothermic tetrapods are significantly different from those influencing bird or rodent communities.

My synthesis is restricted to anurans (frogs and toads) and squamates (lizards, amphisbaenians, and snakes). The other groups of amphibians and reptiles have been excluded for various reasons. Salamanders are absent from most of the lowland tropics. No ecological data are available on the fossorial caecilians. The majority of crocodilians and turtles are aquatic and have little, if any, interactions with other reptiles or amphibians, except as predators.

Data are available from only a few sites for quantitative analysis of the component species of anurans and lizards, and there are comparatively fewer data available on snakes. The most complete study on a lowland rainforest herpetofauna is that by Duellman (1978) on a site (Santa Cecilia) in Amazonian Ecuador. Crump (1971) gave a thorough microhabitat analysis of anurans and lizards at a site near Belém, Brazil. Several studies (Scott, 1976, 1982; Toft, 1980a, 1980b, 1982) have been concerned only with the species inhabiting the leaf litter on the forest floor. Although the diets of many of the species have been documented, only Toft (1980a, 1980b) presented data on the abundance of food items in the environment.

A thorough discussion of species diversity in reptiles was presented by Pianka (1977). Toft (1985) summarized studies of resource partitioning in amphibians and reptiles, and a general synthesis of community ecology of amphibians was presented by Duellman and Trueb (1986).

Comparison of Neotropical Communities

In order to ascertain the species composition of anuran amphibians and squamate reptiles and their utilization of resources, comparative data were synthesized from seven sites in lowland tropical rainforests in Central and South America (Figure 3.1). These sites and the authorities for the data on their herpetofaunas are:

1. Chinajá, Departamento Alta Verapaz, Guatemala—16°02′N, 90°13′W, 140 m elevation (Duellman, 1963). Composition: 14 anurans, 20 lizards, 25 snakes = 59 species.
2. Barro Colorado Island, Canal Zone, Panama—9°10′N, 79°50′W, 150 m elevation (Myers and Rand, 1969). Composition: 27 anurans, 19 lizards and amphisbaenians, 39 snakes = 85 species.
3. Belém, Estado Pará, Brazil—01°21′S, 48°30′W, 12 m elevation (Crump, 1971). Composition: 36 anurans, 26 lizards and amphisbaenians, 45 snakes = 107 species.
4. Santa Cecilia, Provincia Napo, Ecuador—00°03′N, 76°59′W, 340 m elevation (Duellman, 1978). Composition: 86 anurans, 28 lizards and amphisbaenians, 54 snakes = 168 species.
5. Iquitos, Departamento Loreto, Peru—03°43′S, 73°14′W, 105 m elevation; reptiles only (Dixon and Soini, 1975, 1977). Composition: 40 lizards and amphisbaenians, 85 snakes = 125 species of reptiles.

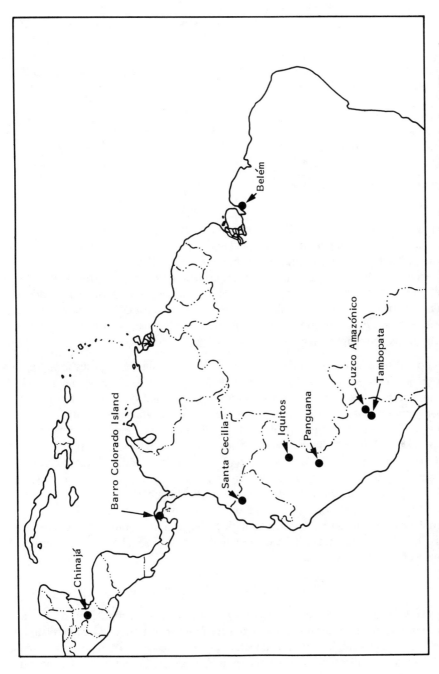

Figure 3.1. Location of sites used in the analysis of lowland tropical rainforest herpetofaunal communities in the Neotropics.

Table 3.1. Major Habitat Utilization and Diel Activity of Species of Squamate Reptiles and Anuran Amphibians at Six Sites in Neotropical Lowland Rainforest

Category	CHINAJA	BA COLO	BELEM	STA CEC	IQ/PAN	MA DIOS
Diurnal						
Arboreal	15 (22)	17 (17)	16 (13)	16 (8)	27 (13)	17 (11)
Terrestrial	30 (44)	45 (43)	46 (37)	70 (35)	96 (44)	55 (34)
Aquatic	0 (0)	0 (0)	2 (1)	1 (1)	0 (0)	0 (0)
Nocturnal						
Arboreal	14 (21)	27 (26)	32 (27)	69 (35)	52 (24)	51 (32)
Terrestrial	8 (12)	15 (14)	22 (18)	38 (19)	31 (14)	32 (20)
Aquatic	1 (1)	0 (0)	6 (4)	5 (2)	10 (5)	5 (3)
Total	68	104	124	199	216	160

Numbers in parentheses are percentages of total number of species.

6. Panguana, Río Llullapichis, Departamento Huánuco, Peru — 09°35'S, 74°48'W, 200 m elevation; amphibians only (Toft and Duellman, 1979; Schlüter 1984). Composition: 62 anurans = 62 species of amphibians.
7. Cuzco Amazónico, Departamento Madre de Dios, Peru — 12°33'S, 69°03'W, 200 m elevation (W.E. Duellman, unpublished data); Tambopata, Departamento Madre de Dios, Peru — 12°50'S, 69°17'W, 290 m elevation (R.W. McDiarmid, personal communication); these two sites combined as Madre de Dios. Composition: 67 anurans, 26 lizards and amphisbaenians, 47 snakes = 140 species.

For purposes of analysis of the entire herpetofauna, the data on reptiles from Iquitos and on amphibians from Panguana were combined into a total herpetofauna of 187 species; both of these two sites are in the same seasonal type of rainforest in the upper Amazon Basin. For reptiles, each species was characterized with respect to habitat utilization (10 categories), diel activity (3 categories), and food (15 categories) (Appendix 1). For amphibians, each species was categorized with respect to habitat utilization (7 categories), diel activity (3 categories), food (5 categories), and reproductive mode (11 categories) (Appendix 2). Species that occupy more than one major habitat type, are diurnal and nocturnal, or feed on more than one group of organisms were categorized in each combination. For example, the viper *Bothrops atrox* is active by day and night on the ground and in bushes and trees; furthermore it feeds on frogs, lizards, and mammals. Thus, the number of "ecological species" exceeds the number of actual species at each site.

Habitat Utilization and Diel Activity

Anyone who has carried out field studies on amphibians and reptiles in neotropical rainforests is aware that the different species are found in different major microhabitats (trees, bushes, ground, leaf litter, and water) and that the species composition in these microhabitats is radically different between night and day. Therefore, it is meaningful to discuss habitat utilization only with respect to diel activity.

Data on these two parameters for the squamate reptiles and anuran amphibians at six sites are summarized in Table 3.1 and illustrated in Figure 3.2. The percentage utiliza-

HERPETOFAUNA

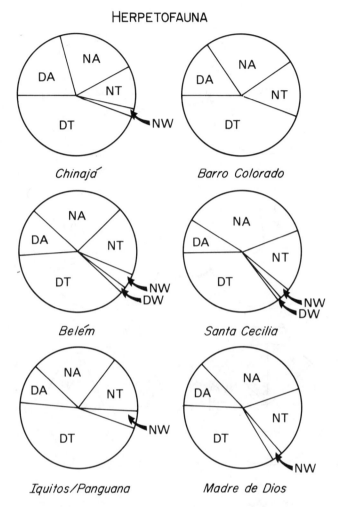

Figure 3.2. Percentage habitat utilization and diel activity of the herpetofauna at six sites. DA = diurnal arboreal, DT = diurnal terrestrial, DW = diurnal aquatic, NA = nocturnal arboreal, NT = nocturnal terrestrial, NW = nocturnal aquatic.

tion of major habitats by day and night by snakes, lizards, and anurans is shown in Figures 3.3–3.5, respectively. From these data, several points are obvious:

1. Approximately half (41–66%) of the species in the herpetofauna at each site are diurnal.
2. Approximately 40% (37–43%) of the species at each site are arboreal.
3. No more than 44% of the herpetofauna is in any one category at each site.
4. Few species (0–5%) are aquatic.
5. The pattern in snakes approximates that of the entire herpetofauna.
6. Generally, more than 50% of the lizards at each site are diurnal and terrestrial.

SNAKES

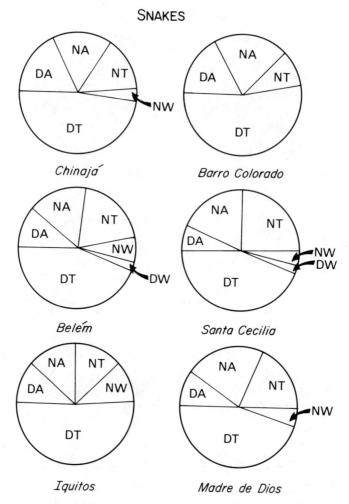

Figure 3.3. Percentage habitat utilization and diel activity of snakes at six sites. Abbreviations as in Figure 3.2.

7. No lizards are nocturnal and terrestrial, and only one or two species of nocturnal arboreal lizards occur at any given site.
8. The majority of all anurans are nocturnal and arboreal, but there are no diurnal arboreal anurans.
9. Generally, there are more nocturnal terrestrial anurans than diurnal terrestrial anurans.

Within each of these broad habitat categories, there are differences in microhabitat utilization and sizes of the species; furthermore, some diurnal speces are heliophilic and others are active only in deep shade. For example, among the nine species of diurnal arboreal lizards at Santa Cecilia, both heliophilic and shade-loving species inhabit

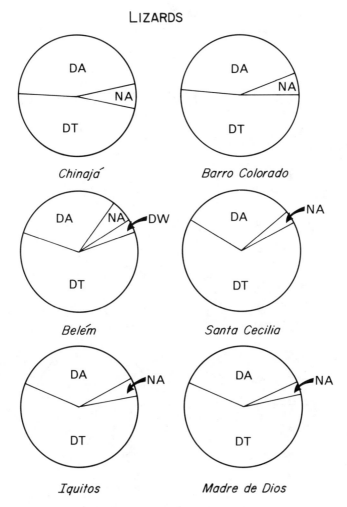

Figure 3.4. Percentage habitat utilization and diel activity of lizards at six sites. Abbreviations as in Figure 3.2.

bushes, tree buttresses, tree trunks, and tree limbs. No more than four species occur in a given category (Figure 3.6), and of these the range in snout-vent lengths is 47–90 mm. Likewise, the 23 species of terrestrial lizards at Santa Cecilia may be grouped into heliophilic and shade species that are fossorial or inhabit open ground, leaf litter, or edges of water courses. The two most abundant subgroups are those that inhabit open ground and are heliophilic and those that inhabit leaf litter in the shade. Of the five species in the former category the range in snout-vent lengths is 50–307 mm, and of the five species in the latter category, 27–70 mm. Likewise, the large assemblages of diurnal terrestrial snakes at each of the sites fall into four groups with respect to shade versus sun and open ground versus leaf litter (Table 3.2).

ANURANS

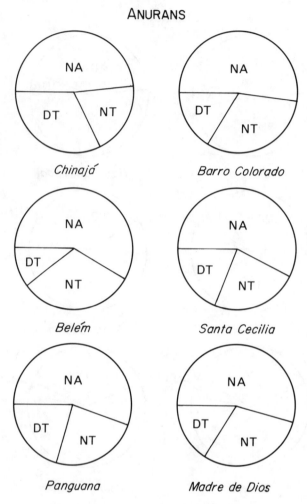

Figure 3.5. Percentage habitat utilization and diel activity of anurans at six sites. Abbreviations as in Figure 3.2.

Utilization of Food Resources

The majority of lizards and anurans are insectivorous; most are generalists, feeding on a wide array of available insects up to a maximum size dictated by the predator's gape. Several species of anurans specialize on ants and a few on termites, whereas ants are the principal (or only) item in the diets of a few species of lizards.

Many species of anurans that are active on the forest floor by day feed on ants. Some species feed primarily or exclusively on ants; the range in size of these frogs is equivalent to that of the feeding generalists, as illustrated by data presented by Duellman (1978) for Santa Cecilia and by Toft (1980a) for Panguana (Figure 3.7). Most species of microhylid frogs feed exclusively on ants on the forest floor at night; at the same time

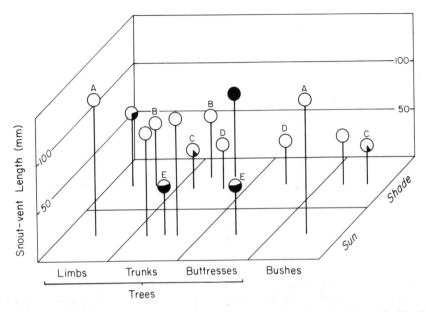

Figure 3.6. Microhabitat utilization by the nine species of arboreal lizards at Santa Cecilia, Ecuador. Species occupying more than one cell are indicated by letters A–E. The percentage of ants in the diet of each species is indicated by black in the circles. Vertical lines to the middle of each circle indicate maximum snout-vent length.

leptodactylid frogs of the genus *Physalaemus* are feeding exclusively on termites, and other leptodactylid frogs are taking a variety of insects. The size range of nocturnal anurans on the forest floor is much greater than that by day, and consequently a greater variety of food is eaten. Ants are eaten primarily by smaller frogs, whereas larger insects, other invertebrates, and even frogs are common in the diets of larger species (Table 3.3).

At each site there is an abundance of insectivorous lizards and at least some anurans that are active by day on the forest floor or amidst leaf litter. The number of species varies from 13 to 37 and accounts for 28–38% of the anuran and lizard fauna at each site. In contrast to numerous species of diurnal anurans that are ant specialists, none of the diurnal terrestrial lizards specializes on ants. The lizards are active foragers and seem to feed on whatever they find that is not too large to overpower and swallow. One of the largest diurnal terrestrial lizards is *Ameiva ameiva*, which forages actively by

Table 3.2. Microhabitat Distribution and Sizes of Terrestrial Diurnal Snakes at Iquitos, Peru

Habitat	No. Species	Range in Total Length (mm)
Open ground, sun	27	247–2270
Open ground, shade	18	265–1318
Leaf litter, sun	2	370–532
Leaf litter, shade	8	251–882

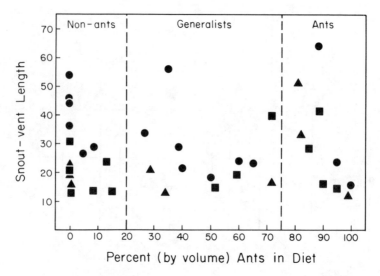

Figure 3.7. Diets and sizes of anurans active by day on the forest floor. Circles = Santa Cecilia throughout the year, squares = Panguana in the wet season, triangles = Panguana in the dry season.

scratching among leaves and turning over small sticks; its diet consists of everything from small isopods to large orthopterans (Duellman, 1978).

Some anurans are active foragers and others are sit-and-wait strategists. Ant specialists are active foragers, whereas non-ant specialists are sit-and-wait strategists, a category that also includes some feeding generalists. With the exception of the ant specialists, most nocturnal anurans seem to be sit-and-wait strategists. Toft (1981) emphasized that diurnal anurans that forage actively on the ground and feed primarily on ants have skin toxins (e.g., *Dendrobates* and *Bufo*) and may be aposematically colored, whereas sit-and-wait strategists are cryptically colored. Furthermore, sit-and-wait strategists are capable of high anaerobic metabolism; they are capable of large bursts of energy for capturing prey or escaping predation. On the other hand, actively foraging species are capable of high aerobic metabolism; they tend to maintain constant but relatively low levels of activity (Bennett, 1982; Duellman and Trueb, 1986).

In contrast, most snakes are feeding specialists. Adults of very few species feed on insects. Some specialize on elongate prey, such as earthworms, caecilians, snakes, and even synbranchid eels. One group, *Dipsas*, specializes on tree snails. However, some neotropical snakes are euryphagous. For example, Seib (1985) reported that the diet of the small, terrestrial snake *Coniophanes fissidens* consisted of salamanders, anurans, anuran eggs, lizards, snakes, reptile eggs, earthworms, and lepidopteran larvae. Some species of pit vipers (e.g., *Bothrops asper* and *B. schlegeli*) that are diurnal and nocturnal are dietary generalists; they feed on anurans, lizards, birds, and small mammals (Greene, 1988).

In lowland rainforests, frogs and lizards are the most common prey items for many species of snakes, whereas some of the larger species feed on small mammals, and a few

Table 3.3. Food Resources of Different Sizes of Nocturnal Terrestrial Anurans at Santa Cecilia, Ecuador

Snout vent Length (mm)	N	Ants	Termites	Other Insects	Spiders	Other Inverts.	Frogs
<50	7	57	14	23	4	2	0
50–100	6	9	0	77	2	11	1
>100	4	9	0	56	9	9	17

Numbers are percentages for N species.

also eat birds. Both diurnal and nocturnal snakes feed on lizards; some nocturnal snakes feed exclusively on diurnal lizards which are preyed upon while they are sleeping at night. Likewise, several diurnal snakes ferret out nocturnal frogs from their diurnal retreats. Thus, the breadth of predation by snakes at any one site encompasses most groups of prey by day and night as shown by the data from Santa Cecilia (Figure 3.8) and Iquitos (Henderson et al., 1979).

Anuran Reproductive Modes

Anurans display a great variety of reproductive modes ranging from the generalized mode of eggs and tadpoles in quite (lentic) water to terrestrial eggs that undergo direct development with no free-living tadpole stage (Duellman and Trueb, 1986). These modes have been analyzed thoroughly at Santa Cecilia (Crump, 1974), and the various modes have been identified for most of the species at the six sites compared in this study (Appendix 2). Species with different reproductive modes utilize different microhabitats for (1) male advertisement calls, (2) oviposition, and (3) larval development.

The male advertisement call is distinctive for each species within a given community. Duellman and Pyles (1983) analyzed call parameters of pond-breeding hylid frogs at three sites (including Belém and Santa Cecilia) and suggested that each acoustic environment is partitioned distinctly by the particular anuran assemblage, and that variation in acoustic partitioning is similar in disjunct communities of like habitat with different species composition.

Data on sites of oviposition and larval development indicate that in most assemblages of lowland tropical forest anurans, deposition of eggs and development of larvae in ponds is the most common mode, but at each site some species have either terrestrial or arboreal eggs that hatch into tadpoles that develop in ponds or streams, and also species which have terrestrial eggs that undergo direct development (Table 3.4).

The most common larval type at all sites is the lentic tadpole; more than 50 such kinds of tadpoles occur at Santa Cecilia and Madre de Dios. Some of these develop in constrained bodies of water, such as water-filled depressions in logs (*Dendrobates*) or in mud basins constructed by adult males (*Hyla boans*). Tadpoles of a few species develop in permanent ponds but the majority develop in temporary ponds. Tadpole communities differ temporally within a pond and spatially between ponds. Furthermore, there are distinct habitat preferences among species of tadpoles in a given pond. Some are pelagic near the surface or in mid-water depths; others are on or near the bottom, and some inhabit only vegetation-choked areas of ponds. For example, among

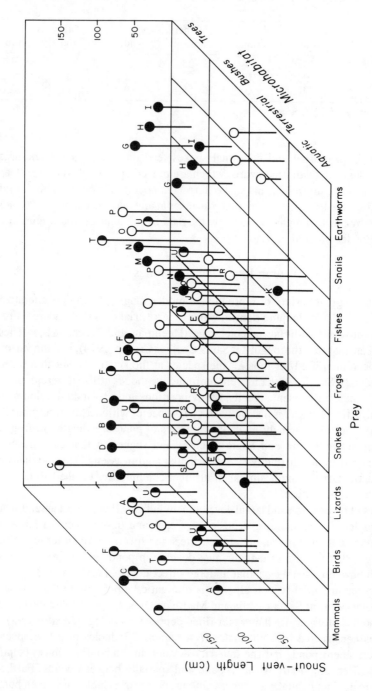

Figure 3.8. Microhabitat utilization, diel activity, and diets of ● = diurnal species, black circles = nocturnal species, half black species of snakes at Santa Cecilia, Ecuador. Species occupying circles = diurnal and nocturnal species, vertical lines to the mid-more than one cell are indicated by the letters A–U. Open circles dle of each circle indicate maximum snout-vent length.

Table 3.4. Anuran Reproductive Modes at Six Sites

Reproductive Mode	CHINAJA	BA COLO	BELEM	STA CEC	PANG	MA DIOS
Lentic eggs:						
Lentic tadpoles	65	22	43	37	35	50
Constrained basin	–	–	–	3	1	1
Lotic eggs:						
Lotic tadpoles	–	7	3	1	1	1
Foam nest:						
Lentic tadpoles	–	11	21	10	12	17
Direct development	–	–	3	1	3	3
Arboreal eggs:						
Lentic tadpoles	14	11	15	12	12	12
Lotic tadpoles	–	11	–	4	3	–
Terrestrial eggs:						
Lentic tadpoles	–	4	9	4	6	5
Lotic tadpoles	–	4	–	3	3	1
Direct development	21	30	3	22	24	9
Brooded eggs:						
Direct development	–	–	3	3	–	1

Numbers are percentages of anurans at each site having the designated reproductive mode.

the 44 species of tadpoles inhabiting temporary ponds at Santa Cecilia, no more than 10 species were found in any given pond at the same time, and these were segregated spatially in the pond (Duellman, 1978).

Species Richness and Community Structure

Examination of Figures 3.2–3.5 reveals that although the number of species inhabiting each of the study sites is highly variable, the percentage utilization of different microhabitats by day and night by the herpetofaunal components is roughly the same at each site. But what are the reasons for the differences?

In some respects when looking at overall diversity, it is not reasonable to include Chinajá and Barro Colorado in the comparison with the other sites, all of which are in the Amazon Basin (see following section on history). Furthermore, Barro Colorado Island was formed by the damming of the Río Chagres, and it was logged in the past; historically, it represents a hilltop on which there are few places for pond-breeding frogs to reproduce. Consequently, some species that are abundant at nearby sites where the terrain is more level are absent on the island.

Among the four sites in the Amazon Basin, Santa Cecilia is unique in having no appreciable dry season, whereas dry seasons having less than 50 mm of rainfall per month last 3–6 mo at the other three sites. The seasonality of rainfall is reflected in the vegetation. At Santa Cecilia, the forest consists of the classic three layers with a nearly continuous canopy (Richards, 1964); throughout the year humidity is high in the lower stratum of the forest and on the ground. At the other sites, there is no continuous canopy, and the amount of humidity in the lower stratum and on the ground varies seasonally. Furthermore, at Iquitos/Panguana and Madre de Dios, the forest is subjected to seasonal flooding. At Belém, there are three types of forest: the terra firme forest never is flooded; the igapó forest is flooded continuously, and the varzea forest is flooded twice daily by tidal effects on the lower Rio Amazonas.

In areas of greater climatic stability (e.g., Santa Cecilia), the forest has a nearly continuous canopy, and there is a layering of forest trees providing a greater diversity of arboreal microhabitats than in seasonally dry forests (e.g., Belém). Moreover, continuously high rainfall and the presence of a nearly continuous canopy result in the greenhouse effect of maintaining high humidity in the lower stories and on the ground (Richards, 1964). Also, the continuously high rainfall results in greater plant productivity which provides an abundance of food for primary consumers, most of which are insects, which in turn are the principal prey of most anurans and lizards.

Climatic stability results in environmental predictability. This is significant to amphibians and reptiles because it assures them of uninterrupted shelter, feeding sites, and food. Furthermore, adequate sites for oviposition and for ovarian and/or larval development are always available. Anurans that deposit their eggs out of water have a high fidelity to forests (Duellman, 1982). High humidity is requisite for the survival of terrestrial or arboreal eggs, and the greatest diversity of anurans having nonaquatic eggs is in areas of continuously high humidity. For example, at Santa Cecilia there are 38 species that deposit nonaquatic eggs; this number diminishes in more seasonal forests to 28 at Panguana, 17 at Madre de Dios, and only 8 at Belém. By comparison, only one such species inhabits the savanna regions of central Venezuela (Hoogmoed and Gorzula, 1979).

Centrolenid frogs have arboreal eggs, and the tadpoles develop in shallow, usually fast-flowing streams. Such streams are absent at Chinajá, Belém, and Madre de Dios, so no centrolenids occur there. However, three species occur at Barro Colorado and Santa Cecilia, and two are found at Panguana. Therefore, even if conditions are optimal for the survival of adults, the species may be absent because of conditions necessary for the development of the eggs or survival of the larvae.

History and Community Composition

Too often community ecologists fail to consider the historical factors influencing the assemblage of species composing a given community. This aspect of community analysis must consider the evolutionary histories of the component species as well as the history of the area of study.

Historically, the biotas of the two Central American sites (Chinajá and Barro Colorado) were independent of those in the Amazon Basin for most of the Tertiary. The herpetofaunas at these two sites contain numerous taxa (e.g., lizards of the genera *Basiliscus*, *Corytophanes*, *Laemanctus*, *Eumeces*, and *Lepidophyma*; frogs of the genera *Agalychnis*, *Smilisca*, and *Syrrhophus*) that evolved in North America or Central America and are not a part of the South American fauna (except in some cases of Quaternary dispersal). Many taxonomic groups (e.g., dendrobatid frogs and microteiid lizards) characteristic of the South American herpetofauna are absent at Chinajá; these groups have not been able to disperse that far north since the closure of the Panamanian Portal in the late Pliocene (Savage, 1982).

The idea of the immutability and antiquity of the Amazonian rainforests was challenged by Haffer (1969), who proposed that climatic changes during the Quaternary resulted in the fragmentation of the rainforests into forest refuges during glacial times. Support for this theory has been provided by a variety of geological and biological studies (Prance, 1982), including some on lizards (Vanzolini and Williams, 1970) and anurans (Duellman and Crump, 1974; Duellman, 1982). The latter proposed a vicariance model in concert with Quaternary climatic-ecological fluctuations to explain the diversity of sympatric, related species in the Amazon Basin.

Theoretically, diversity should be highest in areas of, or adjacent to, Quaternary forest refuges. This assumes that species are still in the process of dispersing from refugia. Santa Cecilia and Iquitos-Panguana have the largest numbers of species; these sites are adjacent to the proposed Napo and Ucayali refuges, respectively (Haffer, 1974).

Discussion

It is obvious that differences exist in the species composition and community structure at different sites in the neotropical rainforests. Some of the differences in species composition and the concomitant participation of these species in the structure of a given community depend on the history of the particular community.

The relative importance of climatic stability and habitat heterogeneity as correlates of diversity has not been determined statistically among rainforest communities. However, Lee (1980) subjected vegetation and climatic data from the Yucatan Peninsula to multivariate analyses. He concluded that in this area, which includes lowland rainforest (Chinajá), tropical deciduous forest, and scrub forest, parameters of vegetation heterogeneity are important correlates of species diversity in lizards and snakes, whereas for amphibians the amount and seasonality of precipitation are the most important.

In examining latitudinal gradients in species diversity of snakes and their prey, Arnold (1972) showed that in areas where there are more species of prey, more species of snakes coexist. Moreover, Henderson et al. (1978) demonstrated a correlation between anuran activity and seasonal incidence of frog-eating snakes at Iquitos.

Whether viewing the total herpetofaunas or their major components (anurans, lizards, snakes), it is evident that in areas of higher climatic stability and vegetational heterogeneity there is a greater number of species than in less stable and less heterogeneous areas. Stability of climates and vegetation creates a stable environment for animals and allows them to specialize on food and microhabitat. Thus, regions with stable climates permit the evolution of finer specializations and adaptations than do regions with more erratic climates because of the relative constancy of resources. Klopfer and MacArthur (1961) theorized that because more species can occupy a given unit of habitat space in stable environments, niches are smaller (i.e., the organisms are more specialized) than in unstable environments. This idea led to the concept of "species packing" in aseasonal tropical environments (MacArthur, 1969, 1970), which has affected the design of subsequent competition and equilibrium models.

The competition models that pervaded community ecology for two decades have been scrutinized critically (see Connor and Simberloff, 1986, for review). As early as 1975, Connell emphasized the importance of predation in determining species composition and in preventing competitive exclusion, especially in benign environments.

Holt (1984) pointed out that instead of considering competition and predation to be alternate, incommensurable mechanisms for community structure, it would be more appropriate to view these interactions more broadly as complementary mechanisms in a unified theory of the ecological niche.

How do competition and predation interact in the structuring of tropical communities of amphibians and reptiles, and how are these factors affected by the environment? In peripheral rainforests having a lengthy dry season, abiotic factors, especially seasonal drought, have detrimental effects on populations of some species, especially anurans. For example, even during the rainy season at Cuzco Amazónico ponds were observed to dry up after a few rainless days; this resulted in the loss of entire cohorts of tadpoles of several species. Aperiodic dry spells even in a normally wet environment such as at Santa Cecilia result in the desiccation of ephemeral ponds that are the breeding sites for many species of frogs (Duellman, 1978).

Predation on anurans, particularly by snakes and birds, is especially high at the time of metamorphosis. Loss of entire cohorts of tadpoles owing to desiccation of ponds will affect predators adversely because they must either find another food resource or starve. On the other hand, the continuous reproduction and metamorphosis of anurans that characterizes environmentally predictable rainforests provides a continuous supply of prey for frog-eating snakes. Of course, not all frog-eating snakes feed solely on metamorphosing young frogs. Some snakes forage at breeding sites at night and capture breeding adult frogs; others forage during the day and capture frogs in their diurnal retreats. Likewise, snakes feed on active lizards by day and sleeping lizards at night.

In less stable rainforest habitats, environmental fluctuations (especially lack of sufficient rainfall) are probably the most important factors reducing population sizes below limits set by resources of many species of anurans and some species of lizards. For example, *Anolis limifrons* is the most common lizard on Barro Colorado Island; however, after an abnormally long dry season, few individuals remain (Andrews and Rand, 1983). The effects of drought may be direct, as they are on larval amphibians, or indirect in diminishing the diversity and abundance of insect prey of frogs and lizards. Diminished food supply can result in interspecific competition. However, Toft (1980a) found that among diurnal terrestrial frogs at Panguana, similarity in diet among feeding guilds tended to be lowest in the dry season when food was less abundant, thereby suggesting not only that was food in shorter supply during the dry season but that the frogs were specializing more and thereby avoiding competition.

The data presented herein represent the best available information on neotropical herpetofaunal communities and only hint at some of the interactions that exist in these communities. There is a desperate need for thorough, long-term studies of such communities in the tropical rainforest. Only by the acquisition of extensive sets of data can we address the major questions and test the multitudinous theories.

Most of the theories concerning tropical rainforest communities have been promulgated by ornithologists. I reiterate that lizards and frogs are not birds. Frogs and lizards are ectothermic feeding generalists, and many of the former have a complex bimodal life-style and are nocturnal. Pough (1980) emphasized that the way of life of amphibians and reptiles is based on low energy flow; their modest energy requirements allow amphibians and reptiles to exploit adaptive zones unavailable to endothermic tetrapods.

We do not expect frogs and lizards to behave like birds, and we should not expect interactions within their communities to be like those of birds.

Lastly, most work on tropical communities has been undertaken by biologists who are native to temperate regions. Upon their first visit to the tropics, these biologists are fascinated by the diversity of species; hence, the oft-repeated question: "Why are there so many species in the tropics?" If we review the climatological history of the earth for the past 55 million years, the great diversity existing in the tropical rainforests is exactly what we should expect. Thus, the more cogent question with respect to the evolutionary aspects of species diversity might be: "Why are there so few species in the temperate zone?"

Acknowledgments. I am indebted to Roy W. McDiarmid for furnishing me a list of species from Tambopata, Peru; Linda S. Dryden and Linda Trueb for executing the illustrations; and Thomas J. Berger, Alan Channing, Harry W. Greene, Robert D. Holt, and Linda Trueb for critically reviewing the manuscript. Data on which this synthesis is based were collected during many years of field work in the American tropics supported at times by the National Geographic Society, National Science Foundation, and the Museum of Natural History, The University of Kansas.

References

Andrews, R.M., and Rand, A.S. 1983. Long term changes in population density of the lizard *Anolis limifrons*. In: Leigh, E.G., Rand, A.S., and Windsor, D.M. (eds.), The Ecology of a Tropical Forest: Seasonal Rhythms and Long-Term Changes, Smithsonian Inst. Press, Washington, D.C., pp. 405–411.

Arnold, S.J. 1972. Species densities of predators and their prey. Am. Nat. 106, 220–236

Bennett, A.F. 1982. The energetics of reptilian activity. In: Gans C., and Pough, F.H. (eds.), Biology of the Reptilia, Academic Press, New York, pp. 155–199

Connell, J.H. 1975. Some mechanisms producing structure in natural communities: A model and evidence from field experiments. In: Cody, M.L., and Diamond, J.M. (eds.), Ecology and Evolution of Communities, Harvard Univ. Press, Cambridge, pp. 460–490

Connor, E.F., and Simberloff, D. 1986. Competition, scientific method, and null models in ecology. Am. Sci. 74, 155–162

Crump, M.L. 1971. Quantitative analysis of the ecological distribution of a tropical herpetofauna. Occas. Pap. Mus. Nat. Hist. Univ. Kansas 3, 1–62

Crump, M.L. 1974. Reproductive strategies in a tropical anuran community. Misc. Publ. Mus. Nat. Hist. Univ. Kansas 61, 1–68

Dixon, J.R., and Soini, P. 1975. The reptiles of the upper Amazon Basin, Iquitos region, Peru. I. Lizards and amphisbaenians. Contr. Biol. Geol. Milwaukee Publ. Mus. 4, 1–58

Dixon, J.R., and Soini, P. 1979. The reptiles of the upper Amazon Basin, Iquitos region, Peru. II. Crocodilians, turtles and snakes. Contr. Biol. Geol. Milwaukee Publ. Mus. 12, 1–91

Duellman, W.E. 1963. Amphibians and reptiles of the rainforests of southern El Petén, Guatemala. Univ. Kansas Publ. Mus. Nat. Hist. 15, 205–249

Duellman, W.E. 1978. The biology of an equatorial herpetofauna in Amazonian Ecuador. Misc. Publ. Mus. Nat. Hist. Univ. Kansas 65, 1–352

Duellman, W.E. 1982. Quaternary climatic-ecological fluctuations in the lowland tropics: Frogs and forests. In: Prance, G.T. (ed.), Biological Diversification in the Tropics, Columbia Univ. Press, New York, pp. 389–402

Duellman, W.E., and Crump, M.L. 1974. Speciation in frogs of the *Hyla parviceps* group in the upper Amazon Basin. Occas. Pap. Mus. Nat. Hist. Univ. Kansas 23, 1–40

Duellman, W.E., and Pyles, R.A. 1983. Acoustic resource partitioning in anuran communities. Copeia 1983, 639–649

Duellman, W.E., and Trueb, L. 1986. Biology of Amphibians. McGraw-Hill, New York

Greene, H.W. 1988. Species richness in tropical predators. In: Almeda, F., and Pringle, C.M. (eds.), Tropical Rainforests: Diversity and Conservation. California Acad. Sci., San Francisco, pp. 259–280

Haffer, J. 1966. Speciation in Amazonian forest birds. Science 165, 131–137

Haffer, J. 1974. Avian speciation in tropical South America. Publ. Nuttal. Ornith. Club 14, 1–390

Henderson, R.W., Dixon, J.R., and Soini, P. 1978. On the seasonal incidence of tropical snakes. Contr. Biol. Geol. Milwaukee Publ. Mus. 17, 1–15

Henderson, R.W., Dixon, J.R., and Soini, P. 1979. Resource partitioning in Amazonian snake communities. Contr. Biol. Geol. Milwaukee Publ. Mus. 22, 1–11

Holt, R.D. 1984. Spatial heterogeneity, indirect interactions, and the coexistence of prey species. Am. Nat. 124, 377–406

Hoogmoed, M.S., and Gorzula, S.J. 1979. Checklist of the savanna inhabiting frogs of the El Manteco region with notes on their ecology and the description of a new species of tree-frog (Hylidae: Anura). Zool. Mededel Leiden 58, 85–115

Klopfer, P.H., and MacArthur, R.H. 1961. On the causes of tropical species diversity: Niche overlap. Am. Nat. 95, 223–226

Lee, J.C. 1980. An ecogeographic analysis of the herpetofauna of the Yucatan Peninsula. Misc. Pub. Mus. Nat. Hist. Univ. Kansas 67, 1–75

MacArthur, R.H. 1969. Species packing and what competition minimizes. Proc. Nat. Acad. Sci. USA 64, 1369–1371

MacArthur, R.H. 1970. Species packing and competitive equilibrium for many species. Theor. Pop. Biol. 1, 1–11

Myers, C.W., and Rand, A.S. 1969. Checklist of amphibians and reptiles of Barro Colorado Island, Panama, with comments on faunal change and sampling. Smithsonian Contr. Zool. 10, 1–11

Pianka, E.R. 1977. Reptilian species diversity. In: Gans, C., and Tinkle, D.W. (eds.), Biology of the Reptilia, Vol. 7, Academic Press, New York, pp. 1–34

Pough, F.H. 1980. The advantages of ectothermy for tetrapods. Am. Nat. 115, 92–112

Prance, G.T. (ed.). 1982. Biological Diversification in the Tropics, Columbia Univ. Press, New York

Rand, A.S., and Humphrey, S.S. 1968. Interspecific competition in the tropical rain forest: Ecological distribution among lizards at Belém, Pará. Proc. U.S. Natl. Mus. 125, 1–17

Richards, P.W. 1964. The Tropical Rain Forest. Cambridge University Press, Cambridge, UK

Savage, J.M. 1982. The enigma of the Central American herpetofauna: Dispersals or vicariance? Ann. Missouri Bot. Gard. 69, 464–547

Schlüter, A. 1984. Ökologische Untersuchungen an einem Stillgewässer im tropischen Regenwald von Peru unter besonderer Berücksichtigung der Amphibien. Ph.D. dissert. Univ. Hamburg

Schoener, T.W. 1974. Resource partitioning in ecological communities. Science 185, 27–39

Scott, N.J., Jr. 1976. The abundance and diversities of the herpetofauna of tropical forest litter. Biotropica 8, 41–58

Scott, N.J., Jr. 1982. The herpetofauna of forest litter plots from Cameroon, Africa. U.S. Fish and Wildlife Res. Rept. 13, 45–150

Seib, R.L. 1985. Euryphagy in a tropical snake, *Coniophanes fissidens*. Biotropica 17, 57–64

Toft, C.A. 1980a. Feeding ecology of thirteen syntopic species of anurans in a seasonal tropical environment. Oecologia 45, 131–141

Toft, C.A. 1980b. Seasonal variation in populations of Panamanian litter frogs and their prey comparison of wetter and dryer sites. Oecologia 47, 34–38

Toft, C.A. 1981. Feeding ecology of Panamanian litter anurans: Patterns in diet and foraging mode. J. Herpetol. 15, 139–144
Toft, C.A. 1982. Community structure of litter anurans in a tropical forest, Makokov, Gabon: A preliminary analysis in the minor dry season. Terre et Vie 36, 223–232
Toft, C.A. 1985. Resource partitioning in amphibians and reptiles. Copeia 1985, 1–21
Toft, C.A., and Duellman, W.E. 1979. Anurans of the lower Rio Llullapichis, Amazonian Peru: A preliminary analysis of community structure. Herpetologica 35, 71–77
Vanzolini, P.E., and Williams, E.E. 1970. South American anoles: The geographic differentiation and evolution of the *Anolis chrysolepis* species group (Sauria, Iguanidae). Arq. Zool São Paulo 19, 1–124

Appendix 1

Distribution of species of squamate reptiles in tropical rainforests at six sites in Central and South America. The sites and authorities for the data are: CH = Chinajá, Departamento Alta Verapaz, Guatemala (Duellman, 1963); BC = Barro Colorado Island, Panama (Myers and Rand, 1969); BE = Belém, Estado Pará, Brazil (Rand and Humphrey, 1968; Crump, 1971; W.E. Duellman, personal data); SC = Santa Cecilia, Provincia Napo, Ecuador (Duellmen, 1978); IQ = Iquitos, Departamento Loreto, Peru (Dixon and Soini, 1975, 1977); MD = Cuzco Amazónico and Tambopata, Departamento Madre de Dios, Peru (W.E. Duellman, personal data; R.W. McDiarmid, personal communication). Habitat utilization is designated by: A = aquatic; AM = aquatic margin; B = bush (< 1.5 m); E = edificarian; F = fossorial; G = ground; L = leaf litter; TB = tree buttresses; TL = tree limbs; TT = tree trunks. Diel activity is designated by: DH = diurnal and heliophylic; DS = diurnal shade; N = nocturnal; ND = nocturnal and diurnal. Food categories are designated: A = primarily or exclusively ants; B = birds; C = caecilians; E = earthworms; F = frogs; G = gastropods; H = herbivorous; I = insects (including spiders and other small arthropods); L = lizards; M = mammals; O = omnivorous; P = fish; S = snakes; T = primarily or exclusively termites; U = salamanders.

Species	CH	BC	BE	SC	IQ	MD	Habitat	Diel	Food
AMPHISBAENIA									
Amphisbaenidae									
Amphisbaena alba	−	−	+	−	+	+	F	DS	I
Amphisbaena fuliginosa	−	+	+	+	+	−	F	DS	E,I
Amphisbaena mitchelli	−	−	+	−	−	−	F	DS	?
SAURIA									
Anguidae									
Celestus rozellae	+	−	−	−	−	−	TT	DS	I
Gekkonidae									
Gonatodes concinnatus	−	−	−	+	+	−	TB,TT	DS	I
Gonatodes hasemani	−	−	−	−	−	+	?	?	?
Gonatodes humeralis	−	−	+	−	+	+	TT	DH	I
Hemidactylus mabouia	−	−	+	−	+	−	E	N	I

Species	CH	BC	BE	SC	IQ	MD	Habitat	Diel	Food
Lepidoblepharus festae	−	−	−	−	+	−	L	DS	I
Lepidoblepharus sanctaemartae	−	+	−	−	−	−	L	DS	I
Pseudogonatodes guianensis	−	−	−	+	+	−	L	DS	I
Sphaerodactylus lineolatus	+	−	−	−	−	−	TT	DH	I
Thecadactylus rapicauda	+	+	+	+	+	+	TT	N	I
Iguanidae									
Anolis biporcatus	+	+	−	−	−	−	TL	DH	I
Anolis bombiceps	−	−	−	−	+	−	G,TT	DS	I
Anolis capito	+	+	−	−	−	−	TT	DH	I
Anolis chrysolepis	−	−	−	+	+	−	G	DS	I
Anolis frenatus	−	+	−	−	−	−	TL	DH	I
Anolis fuscoauratus	−	−	+	+	+	+	G,TT	DS	I
Anolis humilis	+	−	−	−	−	−	G	DS	I
Anolis lemurinus	+	−	−	−	−	−	G,TB	DS	I
Anolis limifrons	+	+	−	−	−	−	G,B	DS	I
Anolis lionotus	−	+	−	−	−	−	AM	DS	I
Anolis ortoni	−	−	+	+	+	+	G,TB,TT	DH	I
Anolis pentaprion	−	+	−	−	−	−	TT	DS	I
Anolis punctatus	−	−	+	+	+	+	TL	DS	I
Anolis sericeus	+	−	−	−	−	−	B	DS	I
Anolis trachyderma	−	−	−	+	+	−	G,B	DS	I
Anolis transversalis	−	−	−	+	+	−	TL,TT	DS	I
Anolis vittigerus	−	+	−	−	−	−	G,TB	DS	I
Anolis sp. A	−	−	−	−	−	+	G	DH	I
Basiliscus basiliscus	−	+	−	−	−	−	AM	DH	I
Basiliscus vittatus	+	−	−	−	−	−	AM,B	DH	I
Corytophanes cristatus	+	+	−	−	−	−	TT	DH	I
Enyalioides cofanorum	−	−	−	+	+	−	G	DH	I
Enyalioides laticeps	−	−	−	+	+	−	TT	DH	I
Enyalioides palpebralis	−	−	−	−	−	+	TT	DH	?
Iguana iguana	+	+	+	−	+	−	AM,TL	DH	H
Laemanctus deborrei	+	−	−	−	−	−	B,TL	DH	I
Ophryoessoides aculeatus	−	−	−	−	+	+	G	DS	I
Plica plica	−	−	−	−	+	+	TT	DH	I
Plica umbra	−	−	+	+	+	+	TT	DS	A
Polychrus gutturosus	−	+	−	−	−	−	B,TL	DH	I
Polychrus marmoratus	−	−	+	+	+	+	B,TL	DH	I
Stenocercus roseiventris	−	−	−	−	−	+	G	DH	I
Tropidurus torquatus	−	−	+	−	−	−	G	DH	I
Uracentron azureum	−	−	−	−	−	+	TT	DH	I
Uracentron flaviceps	−	−	−	+	+	−	TT	DH	I
Uracentron guentheri	−	−	−	−	+	−	TT	DH	I
Uranoscodon supercilaris	−	−	+	−	−	−	AM,TT	DS	I
Scincidae									
Eumeces schwartzei	+	−	−	−	−	−	G	DH	I
Eumeces sumichrasti	+	−	−	−	−	−	G	DH	I
Mabuya mabouya	−	+	+	+	+	+	G	DH	I
Scincella cherrei	+	−	−	−	−	−	L	DS	I
Teiidae									
Alopoglossus angulatus	−	−	−	−	−	+	L	DS	?
Alopoglossus atriventris	−	−	−	+	+	−	L	DS	?
Alopoglossus carinicaudatus	−	−	+	−	+	+	G	DS	?

Species	CH	BC	BE	SC	IQ	MD	Habitat	Diel	Food
Alopoglossus copii	−	−	−	+	−	−	L	DS	I
Alopoglossus sp. A	−	−	+	−	−	−	L	DS	?
Ameiva ameiva	−	−	+	+	+	+	G	DH	I
Ameiva festiva	+	+	−	−	−	−	G	DH	I
Ameiva leptophrys	−	+	−	−	−	−	G	DH	I
Arthrosaura kockii	−	−	+	−	−	−	L	DS	?
Arthrosaura reticulata	−	−	−	+	+	−	L	DS	I
Bachia trisanale	−	−	−	+	+	+	F	DS	E,I
Bachia vermiforme	−	−	−	−	+	−	F	DS	?
Cercosaura ocellata	−	−	−	−	+	+	L	DS	?
Cnemidophorus lemniscatus	−	−	+	−	−	−	G	DH	I
Crocodilurus lacerinus	−	−	+	−	−	−	A	DH	?
Dracaena guianensis	−	−	+	+	+	−	AM	DH	G
Iphisa elegans	−	−	−	+	+	+	L	DS	I
Kentropyx calcaratus	−	−	+	−	+	−	G	DH	I
Kentropyx pelviceps	−	−	−	+	+	+	G	DH	I
Leposoma parietale	−	−	−	+	+	−	L	DH	I
Leposoma percarinatum	−	−	+	−	−	−	AM	DS	I
Leposoma southi	−	+	−	−	−	−	L	DS	I
Neusticurus ecpleopus	−	−	−	+	+	−	AM	DS	I,L
Prionodactylus argulus	−	−	+	+	+	+	L	DH	I
Prionodactylus manicatus	−	−	−	+	−	−	B,L	DH	I
Prionodactylus sp. A	−	−	−	−	−	+	L	DH	?
Tupinambis teguixin	−	−	+	+	+	+	G	DH	O
Xantusiidae									
Lepidophyma flavomaculatum	+	+	−	−	−	−	L	N	I
SERPENTES									
Aniliidae									
Anilius scytale	−	−	+	+	+	−	F	ND	?
Anomalepidae									
Anomalepis dentata	−	+	−	−	−	−	L	DS	?
Liotyphlops albirostris	−	+	−	−	−	−	L	DS	?
Boidae									
Boa constrictor	+	+	−	+	+	+	G,TB	ND	M
Corallus annulatus	−	+	−	−	−	−	TL	N	B
Corallus caninus	−	−	+	+	+	+	TL	N	B,M
Corallus enydris	−	−	−	+	+	+	TL	N	B,M
Epicrates cenchria	−	+	+	+	+	+	G,TL	ND	M
Eunectes murinus	−	−	+	+	+	+	A	ND	M
Colubridae									
Amastridium veliferum	−	+	−	−	−	−	L	DS	F
Apostolepis quinquelineata	−	−	+	−	−	−	L	DS	?
Atractus baudius	−	−	−	−	+	−	G	DS	?
Atractus collaris	−	−	−	−	+	−	G	DS	?
Atractus elaps	−	−	−	+	+	+	G	DS	E
Atractus latifrons	−	−	−	−	+	−	G	DS	?
Atractus major	−	−	−	+	+	+	G	DS	E
Atractus occipitoalbus	−	−	−	+	−	−	G	DS	E
Atractus poeppigi	−	−	−	−	+	−	G	DS	?
Atractus sp. A	−	−	−	−	+	+	G	?	?
Atractus sp. B	−	−	−	−	+	−	G	?	?

Species	CH	BC	BE	SC	IQ	MD	Habitat	Diel	Food
Atractus sp. C	−	−	−	−	+	−	G	?	?
Chironius carinatus	−	+	+	+	+	+	G	DH	F
Chironius exoletus	−	−	−	−	−	+	G	DH	F
Chironius fuscus	−	−	+	+	+	+	G	DH	F,L
Chironius grandisquamus	−	+	−	−	−	−	G	DH	F
Chironius multiventris	−	−	−	+	+	+	G	DH	F
Chironius pyrrhopogon	−	−	−	−	+	−	G	DH	?
Chironius scurrulus	−	−	+	+	+	+	G	DH	F
Clelia bicolor	−	−	−	−	+	−	G	ND	?
Clelia clelia	+	−	+	+	+	+	G	ND	L,M,S
Coniophanes bipunctatus	+	−	−	−	−	−	G	DH	?
Coniophanes fissidens	−	+	−	−	−	−	G	DH	F,L,S
Coniophanes imperialis	+	−	−	−	−	−	G	DH	L
Dendrophidion dendrophis	−	−	−	+	+	−	G	DH	F
Dendrophidion percarinatus	−	+	−	−	−	−	G	DH	?
Dipsas catesbyi	−	−	+	+	+	+	B,TL	N	G
Dipsas indica	−	−	−	+	+	+	B,TL	N	G
Dipsas pavonina	−	−	+	+	+	+	B,TL	N	G
Drepanoides anomalus	−	−	−	+	+	+	G	N	?
Drymarchon corais	+	+	+	−	+	+	G	DH	L,S
Drymobius margaritiferus	+	−	−	−	−	−	G	DH	F
Drymobius rhombifer	−	−	−	+	+	+	G	DH	L
Drymoluber dichrous	−	−	+	+	+	+	G	DH	F,L
Enulius flavitorques	−	+	−	−	−	−	L	DS	?
Enulius sclateri	−	+	−	−	−	−	L	DS	?
Erythrolamprus aesculapii	−	−	+	+	+	+	G	DH	S
Helicops angulatus	−	−	+	+	+	+	A	N	F,L,P
Helicops leopardinus	−	−	−	−	+	+	A	N	F
Helicops pastazae	−	−	−	−	+	−	A	N	?
Helicops petersi	−	−	−	+	−	−	A	N	?
Helicops pelviceps	−	−	+	−	−	−	A	N	?
Helicops polylepis	−	−	−	−	+	−	A	N	?
Helicops trivittatus	−	−	+	−	−	−	A	N	?
Helicops yacu	−	−	−	−	+	−	A	N	?
Hydrops martii	−	−	+	−	+	−	A	N	?
Hydrops triangularis	−	−	+	−	+	−	A	N	?
Imantodes cenchoa	+	+	+	+	+	+	B,TL	N	L
Imantodes gemmistratus	−	+	−	−	−	−	B,TL	N	?
Imantodes lentiferus	−	−	−	+	+	+	B,TL	N	F
Lampropeltis triangulum	+	−	−	−	−	−	G	DH	M
Leptodeira annulata	−	+	+	+	+	+	B,TL	N	F
Leptodeira septentrionalis	+	+	−	−	−	−	B,TL	N	F
Leptophis ahuetulla	−	+	+	+	+	+	B,TL	DH	F
Leptophis cupreus	−	−	−	−	+	−	B,TL	DH	?
Leptophis mexicanus	+	−	−	−	−	−	B,TL	DH	F
Liophis breviceps	−	−	−	−	+	−	G	DH	?
Liophis cobella	−	−	+	+	−	+	G	ND	?
Liophis epinephalus	−	+	−	−	−	−	G	DH	F
Liophis miliaris	−	−	−	+	−	−	G	DS	?
Liophis purpurans	−	−	+	−	−	−	G	DH	?
Liophis reginae	−	−	+	+	+	+	G	DH	F
Liophis typhlus	−	−	−	−	+	−	G	DH	?
Liophis sp. A	−	−	−	+	+	−	G	DH	?

Species	CH	BC	BE	SC	IQ	MD	Habitat	Diel	Food
Liophis sp. B	−	−	−	−	−	+	G,AM	ND	F
Mastigodryas boddaerti	−	−	+	−	−	−	G	DH	?
Mastigodryas melanolomus	+	+	−	−	−	−	G	DH	L
Ninia maculata	−	+	−	−	−	−	L	DS	?
Ninia hudsoni	−	−	−	+	−	−	L	ND	?
Oxybelis aeneus	+	+	−	−	−	−	B,TL	DH	L
Oxybelis argenteus	−	−	+	+	+	+	B,TL	DH	F,L
Oxybelis fuliginosus	−	−	+	−	+	−	TL	DH	B
Oxyrhopus formosus	−	−	−	+	−	−	G	N	L
Oxyrhopus melanogenys	−	−	−	+	+	+	G	ND	L
Oxyrhopus petola	−	+	+	+	+	+	B,G	N	L
Oxyrhopus trigeminus	−	−	+	−	−	−	G	?	?
Philodryas viridissimus	−	−	+	−	+	−	TL	DH	?
Pliocercus euryzonus	+	+	−	−	−	−	G	DH	U
Pseudoboa coronata	−	−	+	+	+	−	G	ND	L,M
Pseudoboa neuwiedii	−	+	−	−	−	−	G	D	?
Pseudoeryx plicatilis	−	−	−	−	+	+	A	N	P
Pseustes poecilonotus	+	+	+	−	+	+	G,B	DH	B,M
Pseustes sulphureus	−	−	−	+	+	+	G,B	DH	M
Rhadinaea brevirostris	−	−	−	+	+	−	L	DH	L
Rhadinaea decorata	−	+	−	−	−	−	L	DH	F,L,U
Rhadinaea occipitalis	−	−	−	−	+	+	L	DH	?
Rhadinaea pachyura	−	+	−	−	−	−	L	DH	U
Sibon dimidiata	+	−	−	−	−	−	B,TL	N	G
Sibon nebulata	+	−	−	−	−	−	B,TL	N	G
Siphlophis cervinus	−	+	+	+	+	+	G	N	L
Spilotes pullatus	+	+	+	−	+	−	G,TL	DH	S
Stenorrhina degenhardtii	+	+	−	−	−	−	G	DH	I
Tantilla albiceps	−	+	−	−	−	−	L	DH	?
Tantilla armillata	−	+	−	−	−	−	L	DH	?
Tantilla melanocephala	−	−	+	+	+	−	L	DH	I
Thamnodynastes pallidus	−	−	−	−	+	−	G,TL	DH	?
Tretanorhinus nigroluteus	+	−	−	−	−	−	A	N	P
Trimetopon barbouri	−	+	−	−	−	−	L	DH	?
Tripanurgos compressus	−	−	+	+	+	−	G,B	N	?
Umbrivaga pygmaea	−	−	−	−	+	−	L	DH	?
Xenodon merremi	−	−	+	−	−	−	G	DH	F
Xenodon rhabdocephalus	+	+	−	−	+	−	G	DH	F
Xenodon severus	−	−	−	+	+	+	G	DH	F
Xenopholis scalaris	−	−	−	+	+	+	L	DS	F
Elapidae									
Micrurus annellatus	−	−	−	−	−	+	L	DS	?
Micrurus diastema	+	−	−	−	−	−	L	DS	S
Micrurus filiformis	−	−	+	−	+	−	L	DS	I
Micrurus hemprichii	−	−	+	−	+	−	L	DS	I
Micrurus langsdorffi	−	−	−	+	+	−	L	DS	S
Micrurus lemniscatus	−	−	+	+	+	−	AM,G	DS	C,P,S
Micrurus mipartitus	−	+	−	−	−	−	L	DS	S
Micrurus narduccii	−	−	−	+	+	−	L	DS	L
Micrurus nigrocinctus	−	+	−	−	−	−	L	DS	?
Micrurus putumayensis	−	−	−	−	+	−	G	DH	S
Micrurus spixii	−	−	+	+	+	+	G	DH	L,S
Micrurus surinamensis	−	−	−	+	+	−	A,AM	DH	P

Species	CH	BC	BE	SC	IQ	MD	Habitat	Diel	Food
Typhlopidae									
Typhlops brongersmianus	−	−	−	−	+	−	L	DS	?
Typhlops minuisquamatus	−	−	−	−	+	−	L	DS	?
Typhlops reticulatus	−	−	+	−	+	+	F,L	DS	?
Viperidae									
Bothrops asper	+	+	−	−	−	−	G	ND	M
Bothrops atrox	−	−	+	+	+	+	G,B	ND	F,L,M
Bothrops bilineatus	−	−	−	+	+	+	B,TL	ND	B,F,M
Bothrops brazili	−	−	−	−	+	−	G	ND	L,M
Bothrops castelnaudi	−	−	−	+	+	−	G	ND	M
Bothrops hypoprorus	−	−	−	−	+	−	G	ND	L,M
Bothrops lichenosus	−	−	+	−	−	−	G	N	?
Bothrops nasutus	+	−	−	−	−	−	G	ND	L,M
Bothrops nummifer	+	−	−	−	−	−	G	ND	M
Bothrops schlegeli	+	+	−	−	−	−	B,TL	ND	M
Lachesis muta	−	−	−	+	+	+	G	N	M

Appendix 2

Distribution of species of anurans in tropical rainforests at six sites in Central and South America. The sites and authorities for the data are CH = Chinajá, Departamento Alta Verapaz, Guatemala (Duellman, 1963); BC = Barro Colorado Island, Panama (Myers and Rand, 1969); BE = Belém, Estado Pará, Brazil (Crump, 1971); SC = Santa Cecilia, Provincia Napo, Ecuador (Duellman, 1978); PA = Panguana, Departmento Huánuco, Peru (Toft and Duellman, 1979; Schlüter, 1984); MD = Cuzco Amazónico and Tambopata, Departamento Madre de Dios, Peru (W.E. Duellman, personal data; R.W. McDiarmid, personal communication). Most species are nocturnal; diurnal species are designated by an asterisk (*). Most species are opportunistic feeders on a variety of small arthropods. The foods of those that are not are noted by the following superscripts: [a]primarily or exclusively ants; f = frogs; p = fish; t = primarily or exclusively termites. Habitat utilization is designated by: A = aquatic; AM = aquatic margin; B = bush; F = fossorial; G = ground; L = leaf litter; T = trees. Reproductive mode is designated by: AELT = arboreal eggs, lentic tadpoles; AEST = arboreal eggs, stream tadpoles; BEDD = brooded eggs on female, direct development; LECB = lentic eggs and tadpoles in constrained basin; LET = lentic eggs and tadpoles; LFLT = lentic foam nest and lentic tadpoles; SET = stream eggs and tadpoles; TEDD = terrestrial eggs, direct development; TELT = terrestrial eggs, lentic tadpoles; TEST = terrestrial eggs, stream tadpoles; TFDD = terrestrial foam nest, direct development.

Species	CH	BC	BE	SC	PA	MD	Habitat	Mode
Bufonidae								
Bufo glaberrimus	−	−	−	+	+	+	G	LET
Bufo marinus	+	+	+	+	+	+	G	LET
Bufo typhonius[*a]	−	+	+	+	+	+	G	SET
Bufo valliceps	+	−	−	−	−	−	G	LET
Dendrophryniscus minutus[*a]	−	−	−	+	+	+	L	LET
Centrolenidae								
Centrolenella fleischmanni	−	+	−	−	−	−	T	AEST
Centrolenella midas	−	−	−	+	+	−	T	AEST
Centrolenella munozorum	−	−	−	+	+	−	T	AEST
Centrolenella prosoblepon	−	+	−	−	−	−	T	AEST
Centrolenella resplendens	−	−	−	+	−	−	T	AEST
Centrolenella spinosa	−	+	−	−	−	−	T	AEST
Dendrobatidae								
Colostethus marchesianus[*a]	−	−	−	+	+	−	L	TEST
Colostethus nubicola[*a]	−	+	−	−	−	−	L	TEST
Colostethus peruvianus[*]	−	−	−	−	+	+	L	TEST
Colostethus sauli[*]	−	−	−	+	−	−	AM	TEST
Dendrobates auratus[*a]	−	+	−	−	−	−	G	TELT
Dendrobates femoralis[*]	−	−	−	+	+	+	G	TELT
Dendrobates parvulus[*a]	−	−	−	+	−	−	G	TELT
Dendrobates petersi[*a]	−	−	−	−	+	−	G	TELT
Dendrobates pictus[*]	−	−	−	+	+	+	G	TELT
Dendrobates quinquevittatus[*]	−	−	+	+	+	−	G	TELT
Dendrobates trivittatus[*]	−	−	+	−	−	+	G	TELT
Hylidae								
Agalychnis calcarifer	−	+	−	−	−	−	T	AELT
Agalychnis callidryas	+	+	−	−	−	−	T	AELT
Agalychnis spurrelli	−	+	−	−	−	−	T	AELT
Hemiphractus proboscideus[f]	−	−	−	+	−	−	L,B	BEDD
Hyla acreana	−	−	−	−	−	+	T	LET
Hyla alboguttata	−	−	−	+	−	−	B	?
Hyla bifurca	−	−	−	+	−	−	T	AELT
Hyla boans	−	−	−	+	+	+	T	LECB
Hyla bokermanni	−	−	−	+	−	−	T	AELT
Hyla brevifrons	−	−	+	+	+	+	T	AELT
Hyla calcarata	−	−	+	+	+	+	T	LET
Hyla ebraccata	+	−	−	−	−	−	T	AELT
Hyla fasciata	−	−	−	+	+	+	T	LET
Hyla geographica	−	−	+	+	+	+	T	LET
Hyla granosa	−	−	+	+	+	+	T	LET
Hyla lanciformis	−	−	−	+	−	+	G,B	LET
Hyla leali	−	−	−	+	−	+	T	LET
Hyla leucophyllata	−	−	+	+	+	+	T	AELT
Hyla loquax	+	−	−	−	−	−	T	LET
Hyla marmorata	−	−	−	+	+	+	T	LET
Hyla melanagyrea	−	−	+	−	−	−	T	LET
Hyla microcephala	+	−	−	−	−	−	T	LET
Hyla minuscula	−	−	+	−	−	−	T	LET
Hyla minuta	−	−	+	+	−	+	T	LET
Hyla multifasciata	−	−	+	−	−	−	G,B	LET
Hyla parviceps	−	−	−	+	+	+	T	LET
Hyla punctata	−	−	−	+	−	+	T	LET

Species	CH	BC	BE	SC	PA	MD	Habitat	Mode
Hyla raniceps	−	−	+	−	−	−	G	LET
Hyla rhodopepla	−	−	−	+	+	+	T	LET
Hyla riveroi	−	−	−	+	+	−	T	LET
Hyla rossalleni	−	−	−	+	−	−	T	?
Hyla rufitela	−	+	−	−	−	−	T	LET
Hyla sarayacuensis	−	−	−	+	+	+	T	AELT
Hyla triangulum	−	−	−	+	+	−	T	AELT
Hyla sp. A	−	−	−	−	−	+	T	LET
Nyctimantis rugiceps	−	−	−	+	−	−	T	LECB
Ololygon boesemani	−	−	+	−	−	−	T	LET
Ololygon boulengeri	−	+	−	−	−	−	T	LET
Ololygon cruentomma	−	−	−	+	+	−	T	LET
Ololygon epacrorhina	−	−	−	−	−	+	T	LET
Ololygon funerea	−	−	−	+	−	−	T	?
Ololygon garbei	−	−	−	+	+	+	T	LET
Ololygon nebulosa	−	−	+	−	−	−	T	?
Ololygon rubra	−	−	+	+	+	+	T	LET
Ololygon staufferi	+	−	−	−	−	−	T	LET
Ololygon x-signata	−	−	+	−	−	−	T	?
Ololygon sp. A	−	−	+	−	−	−	T	?
Osteocephalus buckleyi	−	−	−	+	−	−	T	?
Osteocephalus leprieuri	−	−	−	+	+	+	T	LET
Osteocephalus taurinus	−	−	+	+	+	+	T	LET
Phrynohyas coriacea	−	−	−	+	+	+	T	LET
Phrynohyas venulosa	−	+	+	+	+	+	T	LET
Phyllomedusa atelopoides	−	−	−	−	−	+	G	AELT
Phyllomedusa bicolor	−	−	+	−	−	+	T	AELT
Phyllomedusa hypocondrialis	−	−	+	−	−	−	B	AELT
Phyllomedusa palliata	−	−	−	+	−	+	B	AELT
Phyllomedusa tarsius	−	−	−	+	+	−	T	AELT
Phyllomedusa tomopterna	−	−	−	+	+	+	T	AELT
Phyllomedusa vaillanti	−	−	+	+	+	+	T	AELT
Scarthyla ostinodactyla	−	−	−	−	−	+	T	LET
Smilisca baudinii	+	−	−	−	−	−	T	LET
Smilisca cyanosticta	+	−	−	−	−	−	T	LET
Smilisca phaeota	−	+	−	−	−	−	T	LET
Smilisca sila	−	+	−	−	−	−	T	SET
Sphaenorhynchus carneus[a]	−	−	−	+	−	−	B	?
Sphaenorhynchus lacteus[a]	−	−	+	+	−	+	B	LET
Leptodactylidae								
*Adenomera andreae***[**]	−	−	−	+	+	+	L	TFDD
*Adenomera hylaedactyla***[**]	−	−	−	−	+	+	L	TFDD
*Adenomera marmorata***[**]	−	−	+	−	−	−	L	TFDD
Ceratophrys cornuta[f]	−	−	−	+	+	+	G	LET
*Edalorhina perezi***[*]	−	−	−	+	+	+	L	LFLT
Eleutherodactylus acuminatus[a]	−	−	−	+	+	−	T	TEDD
*Eleutherodactylus altamazonicus***[**]	−	−	−	+	+	−	L,B	TEDD
Eleutherodactylus biporcatus	−	+	−	−	−	−	G	TEDD
Eleutherodactylus bufoniformis	−	+	−	−	−	−	G	TEDD
Eleutherodactylus carvalhoi	−	−	−	−	+	−	L,B	TEDD
Eleutherodactylus caryophyllaceus	−	+	−	−	−	−	L,B	TEDD
*Eleutherodactylus conspicillatus***[**]	−	−	−	+	−	−	G,B	TEDD

Species	CH	BC	BE	SC	PA	MD	Habitat	Mode
Eleutherodactylus crassidigitatus	−	+	−	−	−	−	G,B	TEDD
Eleutherodactylus croceoinguinis	−	−	−	+	−	−	B	TEDD
Eleutherodactylus cruentus	−	+	−	−	−	−	B	TEDD
Eleutherodactylus diadematus	−	−	−	+	+	−	T	TEDD
Eleutherodactylus diastema	−	+	−	−	−	−	B	TEDD
Eleutherodactylus fenestratus	−	−	−	−	−	+	B	TEDD
*Eleutherodactylus gaigeae***	−	+	−	−	−	−	L	TEDD
Eleutherodactylus imitatrix	−	−	−	−	+	+	B	TEDD
Eleutherodactylus lacrimosus	−	−	+	+	+	−	B,T	TEDD
*Eleutherodactylus lanthanites***	−	−	−	+	−	−	G,B	TEDD
Eleutherodactylus martiae	−	−	−	+	−	−	B	TEDD
Eleutherodactylus mendax	−	−	−	−	+	−	B,T	TEDD
*Eleutherodactylus nigrovittatus**	−	−	−	+	−	−	G	TEDD
*Eleutherodactylus ockendeni***	−	−	−	+	−	−	L,B	TEDD
Eleutherodactylus ophnolaimus	−	−	−	+	−	−	T	TEDD
Eleutherodactylus paululus	−	−	−	+	−	−	B	TEDD
Eleutherodactylus peruvianus	−	−	−	−	+	+	G,B	TEDD
Eleutherodactylus pseudoacuminatus	−	−	−	+	−	−	B	TEDD
Eleutherodactylus quaquaversus	−	−	−	+	−	−	B	TEDD
*Eleutherodactylus rostralis**	+	−	−	−	−	−	L	TEDD
*Eleutherodactylus rugulosus**	+	−	−	−	−	−	G	TEDD
*Eleutherodactylus sulcatus**	−	−	−	+	+	−	L	TEDD
Eleutherodactylus taeniatus	−	+	−	−	−	−	B	TEDD
Eleutherodactylus toftae	−	−	−	−	+	+	B	TEDD
Eleutherodactylus variabilis	−	−	−	+	+	−	B	TEDD
Eleutherodactylus ventrimarmoratus	−	−	−	−	+	−	B,T	TEDD
Ischnocnema quixensis	−	−	−	+	+	−	G	TEDD
Leptodactylus bolivianus	−	+	−	−	−	+	G	LFLT
Leptodactylus fuscus	−	−	−	−	−	+	AM	LFLT
Leptodactylus knudseni	−	−	−	−	−	+	G	LFLT
Leptodactylus mystaceus	−	−	+	+	+	+	G	LFLT
Leptodactylus ocellatus	−	−	+	−	−	−	G	LFLT
Leptodactylus pentadactylus	−	+	+	+	+	+	G	LFLT
Leptodactylus podicipinus	−	−	−	−	−	+	G	LFLT
Leptodactylus rhodomystax	−	−	+	+	+	+	G	LFLT
Leptodactylus rhodonotus	−	−	−	−	−	+	G	LFLT
Leptodactylus stenodema	−	−	−	+	−	−	G	LFLT
Leptodactylus wagneri	−	−	+	+	+	+	AM	LFLT
*Lithodytes lineatus***	−	−	−	+	+	+	G	LFLT
Physalaemus ephippifer[t]	−	−	+	−	−	−	G	LFLT
Physalaemus petersi[t]	−	−	+	+	+	+	G	LFLT
Physalaemus pustulosus[t]	−	+	−	−	−	−	G	LFLT
*Syrrhophus leprus***	+	−	−	−	−	−	L	TEDD
*Vanzolinius discodactylus***	−	−	−	+	−	−	AM	LFLT
Microhylidae								
Chiasmocelis anatipes[a]	−	−	−	+	−	−	L	LET
Chiasmocelis bassleri	−	−	−	+	−	−	G	LET
Chiasmocelis ventrimaculata[a]	−	−	−	+	+	+	G	LET
Ctenophryne geayi	−	−	−	+	+	+	G	LET
Elachistocleis ovalis[a]	−	−	−	−	−	+	G	LET
Hamptophryne boliviana[a]	−	−	−	+	+	+	G	LET
Syncope carvalhoi	−	−	−	+	−	−	G	TEDD

Species	CH	BC	BE	SC	PA	MD	Habitat	Mode
Pipidae								
Pipa pipa[p]	−	−	+	+	−	+	A	BEDD
Ranidae								
*Rana berlandieri***	+	−	−	−	−	−	AM	LET
*Rana palmipes***	−	−	−	+	−	−	AM	LET
Rana vaillanti	+	+	−	−	−	−	AM	LET

4. Bird Community Structure in Two Rainforests: Africa (Gabon) and South America (French Guiana) —A Comparison

Christian Erard

Introduction

The high species diversity of tropical rainforests, as compared with temperate woodlands, has attracted a great deal of attention for almost a century. Several recent papers (including Pianka, 1966; Orians, 1969; Terborgh, 1980; Terborgh and Robinson, 1986; Bourlière, 1983, 1984; and Blondel, 1986) have again discussed this issue, and tried to weigh the relative importance of various contributory factors. This is not, in any case, an easy task. Indeed, as pointed out by Bourlière (1984), species richness in any animal community certainly does not depend on a single determinant, but rather on a combination of long-term historical events and shorter term ecological factors.

Historical factors are more and more frequently considered to have played a significant, or even fundamental, role in structuring the communities. In all likelihood, the rate of speciation has always been higher and the rate of extinction probably lower in the humid tropics than in temperate regions (see Haffer, 1969, 1974; MacArthur, 1969; Diamond, 1973; Dorst, 1973, 1974). During the Pleistocene–Holocene the humid tropical regions were affected to a high degree by climatic fluctuations resulting from the effects of glacial periods. These fluctuations and their repercussions on the habitats and on the distribution of species enhanced speciation. But strictly comparable and/or comparative data on speciation and extinction rates in tropical and temperate regions are scanty or even entirely lacking (Vuilleumier, 1969, 1970, 1980, and in litt.). Compared to temperate zones, the tropical regions are credited with a relatively stable climate, little seasonal environmental change, and a certain temporal continuity in the availability of resources (Dobzhansky, 1950; Hutchinson, 1959; Connell and Orias, 1964; Sanders, 1968; MacArthur, 1969, 1972; Diamond, 1978). In the humid tropics resources are supposed be even more varied and abundant, and organic production higher and more sustained than elsewhere (MacArthur, 1958, 1972; Connell and Orias, 1964; Hespenheide, 1971; Schoener, 1971). Thus a finer partitioning of various resource gradients between the different tropical species would be possible and more specialists with narrower realized niches could thus be expected. Tropical species would also tolerate higher degrees of niche overlap, allowing species to be more tightly packed (Klopfer and MacArthur, 1960, 1961; MacArthur et al., 1966; MacArthur and Levins, 1967; Karr, 1971). Tropical communities would also include more guilds (sensu Root 1967) made of a larger number of species distributed along a wider range of sizes (Orians, 1969; Karr, 1971; Schoener, 1971).

It has often been shown that in temperate regions avian species diversity increases with structural diversity of the habitat, expressed in most cases by foliage height diversity (MacArthur and MacArthur, 1961; MacArthur et al., 1962; Recher, 1969; Karr, 1971; Blondel et al., 1973; Roth, 1976). Humid tropical habitats, especially rainforests, are structurally more complex than moist temperate habitats, with a luxuriant vegetation stratified in a number of layers made up of a large number of plant species belonging to a variety of architectural models (sensu Hallé et al., 1978). Such a high structural heterogeneity is expected to enhance, or at least help to maintain, a higher bird species richness (MacArthur et al., 1966; Terborgh, 1977, 1985; Kikkawa, 1982), although some observations do not support this hypothesis (i.e., Orians, 1969; Lovejoy, 1974; Pearson, 1975; Vuilleumier, 1972). It has also been proposed that the humid tropics harbor more predators and/or parasites that contribute to maintain bird populations below their competitive thresholds; in so doing, they would allow more species to share the same habitat (see Pianka, 1966; Paine, 1966).

On the whole, all these explanations are more often based on theoretical reasoning than on hard facts, and they assume that competition plays a fundamental role in structuring communities (see also Vuilleumier, 1979). Although competition does obviously exist, its paramount role has recently been strongly questioned (e.g., Wiens, 1977, 1983; Strong et al., 1979; Connor and Simberloff, 1981; Simberloff, 1984), though other authors maintain a different view (e.g., Schoener, 1983; Gilpin and Diamond, 1984; Grant and Schluter, 1984).

This chapter intends to compare two equatorial forest bird communities, one in Africa (Gabon) and the other in South America (French Guiana),* and to discuss their respective species richness (considered as satisfactorily known; see Vuilleumier, 1978), as a function of various environmental variables. I will avoid the pitfall of comparing diversities on the basis of "bird lists" drawn up after only a few weeks of field work, or on very small plots (e.g., Karr, 1975, 1976a, 1976b, 1980; Karr and James, 1975; Pearson, 1977). My study sites were small enough to minimize the influence of β and γ diversities, and large enough to avoid rare species effects. The present study is complementary to a previous one (Erard, 1986) that was focused on taxonomic problems and less concerned with ecological mechanisms. It is this latter aspect which is developed here with a special emphasis on the guild structure of the two bird commu-

---

*For the sake of convenience French Guiana will subsequently be called Guiana in this chapter.

nities concerned. Results are compared with those of similar studies carried out in South America, especially eastern Peru (Terborgh, 1980, 1985) and Papua New Guinea (Beehler, 1981). Finally, the bird community structure of the two tropical study sites I have personally studied is compared with that of a temperate forest, the pristine forest of Bialowieza, in eastern Poland.

Study Sites

In Gabon investigations were focused on the M'Passa (= Ipassa) plateau near Makokou, at the field station of Institut de Recherches en Ecologie Tropicale (CENAREST). I carried out field work each year during 3–4 consecutive months from 1972 to 1985. André Brosset also made detailed observations on the birds of the area for more than 20 years. Other ornithologists worked for much shorter periods at the same site. The avifauna of the area can therefore be considered as adequately known, and the life history of most species as reasonably well documented, even in detail for some species (Brosset and Erard, 1986; Erard, 1987, and in prep.). The climate is characterized by two rainy seasons (September to November, and February to May), alternating with two dry seasons (June to August, and December and January, the latter being much less marked than the former). Descriptions of the study area have already been published elsewhere (Brosset and Erard, 1986; Caballé, 1978, 1984; Cruiziat, 1966; Florence, 1981; Hladik, 1978, 1986) and need not be repeated here.

In Guiana field work was carried out for 4 consecutive months (in 1980–1981) at Saut Pararé, on the Arataye River, at the end of the dry season and at the beginning of the rains. This was supplemented by a number of visits (totaling 10 months) to other sites between 1980 and 1985. The climate of the area is characterized by a higher annual rainfall than in Gabon (3000 m vs. 1725 mm), by the alternance of a rainy season lasting from December to July (with a brief "dry" spell in February/March) and of a dry season lasting from August to November. Descriptions of the study site are given in the papers of Gasc (1986), Guillotin (1981), Maury-Lechon and Poncy (1986), and Sabatier (1983).

Both the African and the South American study sites are located along a river, but the Ivindo in Gabon is much broader than the Arataye in Guiana. The M'Passa site is also peculiar in that it is surrounded by areas of forest exploited by traditional agriculturalists; it also includes an approximately 20-ha clearing where the field station is located.

Both sites are protected, and the birds were not shy, not being hunted or even disturbed by human activities. At both locations the size of the study area was the same, about 2 km².

Assessment of Species Richness

The only species taken into consideration in the present study are those restricted to the forest environment, i.e., the birds present at the study site because of the forest, and not in spite of it. A forest bird cannot live in open environments, and a nonforest species enters wooded areas only locally and in small numbers (see Amadon, 1973). This distinction is indeed important in order to discriminate between true forest species and those entering the large forest gaps or living on river banks.

In most cases, the birds were identified visually or/and by ear. Each study site was systematically surveyed, the various facies of the forest mosaic being visited at any time of day during each observation period; every observation was recorded on a small-scale map. At both sites the study area was gridded into permanent 100 × 100 m quadrats.

Mist netting allowed me to handle and color-ring most of understorey species, and to check some identifications. Mist netting, however, did not add a single species to the list established by direct observation.

The use of tape recorders to record and play back calls and songs of many species enabled me to observe a number of discrete species that otherwise would have been considered infrequent or even scarce.

These remarks raise the question of potential sampling bias when trying to census rich tropical forest avifaunas (see also Vuilleumier, 1978; Karr, 1981; Wiens, 1983; Terborgh, 1985). Obviously, the value of any estimate of species richness depends both on the size of the study area and on the sampling methods used. For instance, mist netting on too small a surface as compared to the flying range of the birds studied will at best give a very crude estimate of species richness and tell very little, if anything at all, about their relative abundance. Only species dwelling in the lower forest layers where most mist nets are set will also have a good chance of being caught. Because most mist netting and trapping is usually carried out in the lower vegetation layers, the species richness figures have a good chance of being biased toward the species inhabiting the low understorey.

One also has to take into account the variety of social organizations of tropical forest birds (i.e., mixed species foraging associations of insectivores, feeding assemblages of frugivores, or the variety of mating and associated social structures). These social organizations may open unexpected opportunities for exchanging information. In Gabon, for instance, some recently ringed birds were observed literally "warning" other species or their own family members (numerous birds apparently pair for life in Gabon) of the vicinity of the net, or avoiding the regular netting places, even if nets were set at long intervals (Erard, in prep.). The variety of avian social organizations in the tropics, and especially the neotropics, makes it difficult to estimate rainforest bird diversity accurately. Many species live in well-structured social groups, but many others live singly or in pairs or form promiscuous bands. Among the birds living in territorial pairs, it is hard to know what the respective percentages of sedentary breeders and nonbreeders are, whether they are settled or erratic (sensu Smith's 1978 underworld).

Many tropical forest birds have large home ranges. For instance, the home ranges of 14 species of forest flycatchers (Muscicapidae) ranged from 2.4 to 29.5 ha (Erard, 1987). Therefore, in order to capture many species one has to set nets repeatedly on large areas and change the netting sites from time to time in order to avoid habituation phenomena, or active avoidance of mist nets.

In keeping with the social life of tropical forest birds, it is worth pointing out here that more birds appear to hold territories in Gabon than in Guiana, where a large proportion of species form social units which do not seem to be particularly bound to a given forest tract. This difference in territorial behavior has a bearing on the frequency with which some birds vocalize. Fewer species defend territories in Guiana than in Gabon, and birds are less vocally conspicuous and consequently more difficult to detect than in Gabon. Moreover, most South American passerines are suboscines, whereas African passerines are oscines. Both groups differ greatly in their syrinx musculature, so that differences in acoustic behavior can be expected. However, it appears actually very difficult to establish clear relationships between the presence or absence of intrinsic syringeal muscles and the quality of the sounds birds produce (see Brackenbury, 1982).

As a result of all these difficulties no one can presently claim (or should claim) to have ever computed diversity indices based on accurate estimates of the abundances of all species for any tropical bird community, especially in rainforests. Any ecologically significant sampling of a bird community should deal with true densities, i.e., numbers of individuals per unit area, and not breeding pairs per unit area, as is usually done in the ornithological literature. This would make possible better and more meaningful comparisons with other syntopic animal components (other vertebrates in particular) of the ecosystem.

Also relevant to a correct approach of tropical forest bird diversity is the question of how to define satisfactorily the rainforest mosaic which—this should never be forgotten—is a tridimensional mosaic. Comparisons between sites, particularly when these sites are not located on the same continent, should also be based on an objective description of forest microhabitats. Such descriptions should include the distribution and rate of occurrence of recent forest gaps, the apportionment of the different stages of forest regeneration, the frequency of occurrence of the various architectural models of tropical trees (sensu Hallé et al., 1978), and the delimitation of the interface between the euphotic canopy and the oligophotic understorey, based on the localization of the "morphological inversion" zone of the "mature trees" (sensu Oldeman, 1974). I agree with Terborgh (1985) that habitat categorization is still, in most cases, oversimplified by ornithologists.

Taxonomic Composition

This topic has been discussed in detail previously (Erard, 1986) and I need only mention here that 227 forest species (out of a grand total of 292) belonging to 15 orders, 39 families, and 154 genera have been recorded at Saut Pararé, as against 175 species (grand total 364) representing 14 orders, 34 families, and 104 genera at M'Passa. There are about the same numbers of nonpasserines in the two continents (24 families, 57 genera, 88 species in Guiana vs., respectively, 20, 56, and 81 in Gabon), but this is not the case for passerines, except for the number of families (15 families, 97 genera, and 139 species in Guiana vs. 14, 48, and 94 in Gabon).

Some cases of morphological and ecological convergences (or parallelisms due to some taxonomic affinity?) are very striking in the nonpasserines. They include the herons *Tigriornis leucolophus* in Gabon and *Tigrisoma lineatum* in Guiana, the ibises *Bostrychia rara* in Africa and *Mesembrinibis cayennensis* in South America, or the owls *Jubula (Lophostrix) letti* in Gabon and *Lophostrix cristata* in Guiana.

Table 4.1. Distribution of Numbers of Species in the Two Equatorial Rainforest Sites According to Broad Ecological Categories

Ecological Categories	Taxonomic Categories	No. of Species (%) in	
		Gabon	F. Guiana
	Nonpasserines	10 (12.3%)	8 (9.1%)
Aquatic habitats	Passerines	5 (5.3%)	1 (0.7%)
	Total	15 (8.6%)	9 (4.0%)
Carrion feeders	Total (all NP)	–	1 (0.4%)
	Diurnal (all NP)	9 (11.1%)	10 (11.4%)
Raptors	Nocturnal (all NP)	4 (4.9%)	4 (4.5%)
	Total	13 (7.4%)	14 (6.2%)
Terrestrial feeders	Nonpasserines	7 (8.6%)	8 (9.1%)
	Passerines	1 (1.1%)	14 (10.4%)
	Total	8 (4.6%)	22 (9.7%)
Arboreal (+ aerial) feeders	Nonpasserines	51 (63.0%)	57 (64.8%)
	Passerines	88 (93.6%)	124 (89.2%)
	Total	139 (79.4%)	181 (79.7%)
Total avifauna	Nonpasserines	81 (100%)	88 (100%)
	Passerines	94 (100%)	139 (100%)
	Total	175 (100%)	227 (100%)

The two sites share 13 orders out of 16, but only 15 families out of 57 (among them 5 passerine families) and 7 genera out of 251. This reflects a fundamental phylogenetic difference between them. This is not surprising when the past history of isolation of the two continents is taken into consideration (see, for example, Moreau, 1966; Keast, 1972a; Cracraft, 1973).

Gross Ecological Structure

In Table 4.1 I compare the communities of Gabon and Guiana on the basis of broad ecological categories adapted from those defined by Moreau (1966), Hamel (1980), and Short (1980). These categories were modified so that the various component species of a given family could be individually placed in the "right" category instead of being allocated globally to a category that fits some of them but not the others. For instance, the species of Alcedinidae were here divided into "aquatic" species and species found in wooded habitats, in contrast to Moreau (1966), who allocated all kingfishers to aquatic habitats.

The following categories have been recognized: (1) species confined to aquatic habitats (i.e., river banks), (2) carrion eaters, (3) diurnal and nocturnal raptors, (4) birds that are mostly active at ground level, and (5) birds that live in trees or hunt for prey above the forest canopy.

For the whole avifauna, the following differences become readily apparent when the Guiana site is compared to the Gabon site: fewer bird species confined to aquatic

Table 4.2. Species Richness in Relation to Food Type

Food Type	Birds	No. of Species (%) in	
		Gabon	F. Guiana
Aquatic organisms	Nonpasserines	–	10 (11.4%)
	Passerines	–	–
	Total	7 (4.0%)	6 (2.6%)
Nectar	Nonpasserines	–	10 (11.4%)
	Passerines	1 (1.1%)	3 (2.2%)
	Total	1 (0.6%)	13 (5.7%)
Small seeds	Nonpasserines	3 (3.7%)	1 (1.1%)
	Passerines	1 (1.1%)	–
	Total	4 (2.3%)	1 (0.5%)
Fruit	Nonpasserines	15 (18.5%)	34 (38.6%)
	Passerines	5 (5.3%)	32 (74.8%)
	Total	20 (11.4%)	66 (29.1%)
Arthropods	Nonpasserines	45 (55.6%)	23 (26.2%)
	Passerines	87 (92.6%)	104 (74.8%)
	Total	132 (75.4%)	127 (55.9%)
Vertebrates	Nonpasserines	11 (13.6%)	13 (14.8%)
	Passerines	–	–
	Total	11 (6.3%)	13 (5.7%)
Carrion	Total (all NP)	–	1 (0.5%)
Total avifauna	Nonpasserines	81 (100%)	88 (100%)
	Passerines	94 (100%)	139 (100%)
	Total	175 (100%)	227 (100%)

habitats (especially passerines), more ground-living birds (again more passerines), and more arboreal passerines (in absolute values, not in percentages) at the Guiana site. The presence of forest vultures in Guiana, a fact related to the ability of *Cathartes* species to locate carrion by olfaction and not only by sight as Old World vultures do, is also· worth noting. Attention should also be drawn to the absence of neotropical fishing owls in Guiana whereas two species exist in Gabon: their ecological niche seems to be filled in South America by fish-eating bats (*Noctilio*).

Trophic Structure

A comparison based on the dietary habits is given in Table 4.2, where the species are classified according to the following trophic categories: (1) birds feeding mainly on aquatic organisms; (2) nectarivores, (3) small seed-eaters (granivores), (4) frugivores (species whose diet includes more than 60% fruit, no distinction being made between seed dispersers or seed predators), (5) birds feeding on arthropods, (6) vertebrate predators, and (7) carrion eaters.

The figures in Table 4.2 suggest six major differences between the African and South American sites:

(1) The proportion of species confined to aquatic habitats tends to be greater in Gabon than in Guiana, but such a trend is less obvious when only the birds feeding on aquatic organisms are considered. It would thus appear that riverine vegetation provides opportunities for nonaquatic birds to isolate themselves from other syntopic species that are either taxonomiclaly related or ecologically similar.

(2) Nectarivores are more numerous in Guiana than in Gabon, this difference being mainly due to the large number of hummingbirds (Trochilidae) in the neotropics. Particularly worth noting is the role that flowers play as food sources for birds in Guiana (not to mention bats), especially at the end of the long dry season, which is a time of diminishing supply of fruit and insects. Then every flowering tree attracts a number of bird species, which congregate there not only to catch insects but also to feed on nectar and even ingest whole flowers. I can incidentally report here of observing birds (e.g., *Cyanerpes* and *Cacicus*) feeding flowers to their young.

(3) The number of small seed-eaters is low in Guiana.

(4) There are many more frugivorous birds in Guiana than in Gabon. This is true for nonpasserines as well as for passerines, but frugivory is particularly developed among the latter: 5.3 species per family in Guiana against 1.7 in Gabon.

(5) Nonpasserines and passerines are significantly more "insectivorous" in Gabon than in Guiana.

(6) The assemblage of diurnal and nocturnal predators of vertebrates is roughly the same in numbers, but not in kind, at both sites.

Spatial Utilization of the Habitat

It would be premature at present to compare horizontal habitat utilization by birds in the two forest communities studied. Adequate data are available for most Gabonese species but this is not the case for Guiana, where not enough is known about microhabitat preferences, social organization, home range sizes, and even territorial behavior.

Nevertheless a few remarks can be made. First of all, it is important to emphasize that in Gabon as well as in Guiana, long-range migrants never enter the forest. They occur only along its edges, along the largest rivers, and also in second growth vegetation.

In Gabon as in Guiana there are "erratic" bird species in the forest whose range of movements remains unclear. Regular and predictable migrations appear to take place among frugivores, such as parrots and toucans in Guiana. There also seems to be more erratic species in Guiana than in Gabon, where true frugivores are scarcer.

More species also appear to be nomadic in Guiana than in Gabon, though it remains impossible at this time to distinguish species whose social units are centered on a given spot while having very large home ranges from those that spend the whole year in endlessly wandering social units. Representatives of such nomadic birds move over vast areas without displaying any territorial behavior. They can even become gregarious at times, searching for patches of food irregularly distributed in time and in space throughout the forest (such as flowering or fruiting trees). Many hummingbirds, tanagers, and cotingas, among others, may belong to this nomadic category.

As already mentioned, many more bird species appear to be territorial in Gabon than in Guiana. However, the size of territories and the duration of their tenure remain to be established in most cases. The scanty information available supports the view that for a given size and kind of species, the territories are larger in Gabon and Guiana than in temperate forests. Furthermore, mixed species foraging associations of insectivores, with their set of core species (see Munn, 1985; Munn and Terborgh, 1979; Powell 1980, 1985), are frequently observed in Guiana – the associated species having strictly overlapping and jointly defended territories. Such long-lasting and stable interspecific associations have never been observed in Gabon so far; there, territorial limits do not coincide so closely, and the existing foraging associations are apparently not based on such elaborate mutualistic behavior.

Despite our inadequate knowledge of the foraging behavior of most bird species of Guiana, the observations at hand allow a preliminary analysis of the vertical use of space within the forest (see Table 4.3, from which raptors and birds feeding on aquatic organisms were excluded).

It immediately appears that, though there are more species in Guiana (22) than in Gabon (16), the proportion of ground-feeding birds is about the same at both sites (9.7 and 9.2%). However, the proportion of frugivores and animal eaters ("carnivores") among passerines and nonpasserines is the reverse: Guiana harbors more fruit-eating nonpasserines and more animal-eating passerines than Gabon.

A striking point emerging from Table 4.3 is the high species richness in the canopy (>20 m): 50.8% of species in Gabon and 39.6% in Guiana are found at that level. The larger percentage of canopy birds in Gabon is essentially provided by passerines. It also appears that species richness increases from ground level to canopy at both sites, but the occupancy of the different understorey levels is more homogeneous in Guiana than in Gabon; this applies to passerines as well as to nonpasserines. This is particularly obvious when the vertical distribution of foliage arthropod-feeding species is considered.

Vegetarian ("herbivore") species apparently make a better use of the understorey in Guiana, where 4.9% of the frugivores and 3.5% of the nectarivores are found, than in Gabon, where all frugivores live in the canopy. It would indeed be most interesting to quantify the relative importance of cauliflory and caulicarpy among trees at both sites. It appears at first sight that such plants are more numerous in Guiana than in Gabon.

Guild Structure

In order to account for the different kinds of foraging behaviors in the analysis of forest bird community structure, I have used the guild partitioning of the avifaunas proposed by Terborgh (1980) and also followed by Beehler (1981). Because no precise definition of the various guilds were given in these papers, however, I cannot be absolutely certain that my assignments of species to some of Terborgh and Beehler's categories is correct. To avoid any misunderstanding, information on my classification criteria is given below.

Table 4.3. Species Richness in Relation to Main Food Types and Localization During Feeding

Food Type	Foraging Substrate		Gabon			F. Guiana		
			Nonpasserines	Passerines	Total	Nonpasserines	Passerines	Total
Arthropods	Ground level		7 (8.6%)	5 (5.3%)	12 (6.9%)	—	13 (9.4%)	13 (5.7%)
	Trunks	0–2m	2 (2.5%)	1 (1.1%)	3 (1.7%)	3 (3.4%)	5 (3.6%)	8 (3.5%)
		2–10m	1 (1.2%)	2 (2.1%)	3 (1.7%)	4 (4.5%)	6 (4.3%)	10 (4.4%)
		10–20m	2 (2.5%)	1 (1.1%)	3 (1.7%)	4 (4.5%)	7 (5.1%)	11 (4.8%)
		>20m	—	—	—	—	—	—
	Foliage	0–2m	1 (1.2%)	15 (15.9%)	16 (9.1%)	—	16 (11.5%)	16 (7.0%)
		2–10m	2 (2.5%)	6 (6.4%)	8 (4.6%)	1 (1.1%)	20 (14.4%)	21 (11.9%)
		10–20m	3 (3.7%)	17 (18.1%)	20 (11.4%)	4 (4.5%)	17 (12.2%)	21 (11.9%)
		>20m	25 (30.9%)	40 (42.5%)	65 (37.1%)	7 (7.9%)	20 (14.4%)	27 (11.9%)
Seeds + fruits	Ground level		3 (3.7%)	1 (1.1%)	4 (2.3%)	8 (9.1%)	1 (0.7%)	9 (4.0%)
	Trees	0–2m	—	—	—	—	4 (2.9%)	4 (1.8%)
		2–10m	—	—	—	1 (1.1%)	6 (4.3%)	7 (3.1%)
		10–20m	—	—	—	—	—	—
		>20m	15 (18.5%)	5 (5.3%)	20 (11.4%)	26 (29.5%)	21 (15.1%)	47 (20.7%)
Nectar		0–2m	—	—	—	7 (7.9%)	—	7 (3.1%)
		2–10m	—	—	—	—	1 (0.7%)	1 (0.4%)
		10–20m	—	1 (1.1%)	1 (0.6%)	—	—	—
		>20m	—	—	—	3 (3.4%)	2 (1.4%)	5 (2.2%)

I considered as frugivores or as insectivores birds whose diet includes more than 70% of fruits or arthropods, respectively. Otherwise they were placed in mixed categories. The guild of mast eaters is nowhere defined either by Terborgh or by Beehler. We are not told whether a bird must be considered a mast eater because it feeds on a certain type of fruit (e.g., acorns or fruits with a high seed/pulp ratio), or because it specializes on fruits of mast-fruiting trees (e.g., fruits of trees having a highly synchronized, periodic, localized, and heavy fruit production, such as the Southeast Asian Dipterocarps), or for other reasons. Given the great variability that exists in the morphology of fruits and in the pattern of production of fruiting trees in the forests studied (see Gautier-Hion et al., 1985a, 1985b; Sabatier, 1982, 1983, 1985; Sabatier and Puig, 1986), I decided to consider separately the birds that feed on seeds (e.g., pigeons, parrots, tinamous, curassows) and the true frugivorous species that feed on fruit pulps but disperse the seeds. I feel that such a distinction is ecologically important; it allows one to allocate species according to the quality of the fruit they consume (i.e., the kind of reward they receive) and according to the positive or negative role they play on the dispersal of plant diaspores, i.e., their different roles in forest dynamics.

Like Beehler (1981), I followed Terborgh (1980) and placed all hummingbirds and honeycreepers among nectarivores because they all feed essentially on nectar and insects constitute only a small fraction of their diet. I nevertheless realize that such a viewpoint can be disputed, but at our present state of knowledge it would be even more questionable to do otherwise.

Likewise, in the absence of definition of Terborgh's or Beehler's category of frugivores/predators, I included only one species in it: the Falconidae *Daptrius americanus*.

Terborgh (1980) set apart as a category the frugivores/insectivores/nectarivores. I do not think that such a guild exists in Gabon, except perhaps for *Zosterops senegalensis*, a species that has only been recorded once at Bélinga but not at M'Passa (Brosset and Erard, 1986). Our knowledge of the diet of the birds of Guiana is still not good enough to decide whether such a guild is present or not (e.g., in Thraupinae); in secondary successions and other man-modified habitats a species such as *Rhamphocelus carbo*, for instance, could perhaps be included in this mixed diet guild.

Beehler (1981) distinguished a lichen gleaner guild in the New Guinean highland forests (above 1000 m) he studied. I did not find any representative of such a guild either in Guiana or in Gabon, although at Bélinga, in forests located at an altitude of about 1000 m, many trees covered with usnea were visited by several species of sunbirds, whose foraging, however, was not limited to this microhabitat.

Intertropical Comparisons

A comparison of the composition of the Gabonese and Guianan avifaunas classified according to trophic/behavioral guilds (Tables 4.4 and 4.5) suggests a number of interesting differences. First it appears that three nonspecific guilds present in Guiana are lacking in Gabon: the carrion eaters, the terrestrial frugivores, and the frugivores/predators.

Because all the field work was done by myself, as well as the repartition of species in different guilds, the comparison of the emerging patterns can be considered as quite meaningful.

Table 4.4. Guild Structure (Number of Species and Proportion) Relative to the Whole Avifauna of 4 Tropical and 1 Temperate Forest Bird Communities

Trophic/Behavioral Guild	No. of Species (%) in				
	Gabon (N=175)	F. Guiana (N=227)	E. Peru (N=207)	P.N.G. (N=151)	Poland (N=56)
Carrion	—	1 (0.4%)	2 (0.97%)	—	1 (1.8%)
Raptor					
general	5 (2.9%)	7 (3.1%)	7 (3.38%)	5 (3.31%)	2 (3.6%)
bird	3 (1.7%)	1 (0.4%)	4 (1.93%)	5 (3.31%)	2 (3.6%)
other specialist	1 (0.6%)	1 (0.4%)	7 (3.38%)	—	1 (1.8%)
Owl	6 (3.4%)	4 (1.8%)	5 (2.42%)	4 (2.65%)	2 (3.6%)
Nightjar	2 (1.1%)	2 (0.9%)	1 (0.48%)	3 (1.99%)	—
Mast eater					
terrestrial	4 (2.3%)	8 (3.5%)	5 (2.42%)	1 (0.66%)	4 (7.1%)
arboreal	5 (2.9%)	17 (7.5%)	8 (3.86%)	2 (1.32%)	3 (5.4%)
Frugivore					
terrestrial	—	1 (0.4%)	3 (1.45%)	11 (7.28%)	—
arboreal	12 (6.8%)	26 (11.5%)	18 (8.70%)	30 (19.87%)	—
Nectarivore	1 (0.6%)	13 (5.7%)	8 (3.86%)	10 (6.62%)	—

Insectivore					
terrestrial	9 (5.1%)	13 (5.7%)	10 (4.83%)	9 (5.96%)	7 (12.5%)
woodpecker	5 (2.9%)	11 (4.9%)	8 (3.86%)	—	5 (8.9%)
bark-gleaning	4 (2.3%)	16 (7.1%)	9 (4.35%)	3 (1.99%)	2 (3.6%)
foliage-gleaning	72 (41.1%)	41 (18.1%)	19 (9.18%)	16 (10.60%)	17 (30.3%)
sallying	21 (12.0%)	20 (8.8%)	27 (13.04%)	13 (8.61%)	5 (8.9%)
aerial	2 (1.1%)	1 (0.4%)	4 (1.93%)	2 (1.32%)	—
ant-following	4 (2.3%)	7 (3.1%)	6 (2.90%)	—	—
dead leaf-gleaning	1 (0.6%)	11 (4.9%)	7 (3.38%)	—	—
vine-gleaning	4 (2.3%)	5 (2.2%)	7 (3.38%)		1 (1.8%)
Frugivore/predator	—	1 (0.4%)	6 (2.90%)	3 (1.99%)	—
Frugivore/insectivore					
arboreal, gleaning	6 (3.4%)	10 (4.4%)	12 (5.80%)	20 (13.25%)	4 (7.1%)
arboreal, sallying	1 (0.6%)	4 (1.8%)	13 (6.28%)	—	—
Frugivore/insectivore/nectarivore	—	—	11 (5.31%)	6 (3.97%)	—
Lichen-gleaning			—	1 (0.66%)	—
Aquatic	7 (4.0%)	6 (2.6%)	—	2 (1.32%)	—
Arboreal-to-ground insectivore	—	—	—	4 (2.65%)	—

Source: E. Peruvian data from Terborgh (1980); Papuan New Guinean data from Beehler (1981); Polish data from Tomialojc et al. (1984).

Table 4.5. Guild Structure of Two Tropical Forest Bird Communities in Gabon and French Guiana

Trophic/behavioral guild	Nonpasserines		Passerines	
	Gabon (N=81)	F. Guiana (N=88)	Gabon (N=94)	F. Guiana (N=139)
Carrion	—	1 (1.1%)	—	—
Raptor				
general	5 (6.2%)	7 (8.0%)	—	—
bird	3 (3.7%)	1 (1.1%)	—	—
other specialist	1 (1.2%)	1 (1.1%)	—	—
Owl	6 (7.4%)	4 (4.5%)	—	—
Nightjar	2 (2.5%)	2 (2.3%)	—	—
Mast eater				
terrestrial	3 (3.7%)	7 (8.0%)	1 (1.1%)	1 (0.7%)
arboreal	5 (6.2%)	17 (19.3%)	—	—
Frugivore				
terrestrial	—	1 (1.1%)	—	—
aborbeal	10 (12.3%)	7 (8.0%)	2 (2.1%)	19 (13.7%)
Nectarivore	—	10 (11.4%)	1 (1.1%)	3 (2.2%)
Insectivore				
terrestrial	4 (4.9%)	—	5 (5.3%)	13 (9.4%)
woodpecker	5 (6.2%)	11 (12.5%)	—	—
bark-gleaning	16 (19.8%)	—	4 (4.3%)	16 (11.5%)
foliage-gleaning	11 (13.6%)	3 (3.4%)	56 (59.5%)	38 (27.3%)
sallying	—	7 (8.0%)	10 (10.5%)	13 (9.4%)
aerial	—	—	2 (2.1%)	1 (0.7%)
ant-following	—	—	4 (4.3%)	7 (5.0%)
dead leaf-gleaning	—	—	1 (1.1%)	11 (7.9%)
vine-gleaning	—	—	4 (4.3%)	5 (3.6%)
Frugivore/predator	—	1 (1.1%)	—	—
Frugivore/insectivore				
arboreal, gleaning	2 (2.5%)	2 (2.3%)	4 (4.3%)	10 (7.2%)
arboreal, sallying	1 (1.2%)	6 (6.8%)	—	2 (1.4%)
Aquatic	7 (8.6%)	—	—	—

Table 4.6. Simpson's Index, and Standard Diversity Index Calculated from the Distribution of Species in the Different Guilds in the Two Equatorial Forests Studied

	Gabon			F. Guiana		
	N	D	Ds	N	D	Ds
Nonpasserines	15	9.62	0.61	17	9.90	0.55
Passerines	12	2.65	0.15	13	7.13	0.51
Avifauna	21	5.00	0.20	24	12.08	0.48

Note: N = number of guilds present; D = Simpson's diversity index ($D = 1/\Sigma p_i^2$); Ds = standard diversity index [$Ds = (D - 1)/(N - 1)$].

The allocation of the birds to the guild categories in Gabon was compared to the one in Guiana by using Simpson's diversity index (Levins, 1968), $D = 1/\Sigma p_i^2$ (with p_i being the proportion of species in guild i; D varies from 1 when all species are in the same guild to N when species are distributed in the same proportion in the N guilds), and the standard diversity index $D_s = (D - 1)/(N - 1)$ (which varies between 0 and 1). The results are presented in Table 4.6. The most salient point is that the various species of the species pool (and more particularly the passerines) are more equally apportioned among the different guilds in Guiana than they are in Gabon. This is essentially due to the fact that in Gabon many species are insectivores, particularly foliage gleaners.

Apart from the fact that there are more frugivorous species (53 vs. 21 species) and nectarivorous species (13 vs. 1 species) in Guiana than in Gabon, it appears that the proportion of seed predators among the fruit-eating birds of Guiana is higher than in Gabon (11% vs. 5.2%). This is particularly true for the arboreal frugivores (7.5% vs. 2.9%). This is due to the presence of Guiana of a number of large terrestrial species of frugivores and of a much larger number of parrots. Guiana is also richer in arboreal frugivorous passerines able to disperse seeds (19 vs. 2 species in Gabon). The difference is even more spectacular if we group them together with mixed diet species (31 vs. 6 species).

As for the insectivores, it must be emphasized that the difference between the two sites (a higher proportion of insectivores in Gabon than in Guinea) is only relative; if we consider their absolute numbers, the figures are almost the same: 122 species in Gabon, 36 of which are nonpasserines and 86 passerines, vs. 125, 21, and 104, respectively, in Guiana (without including frugivores/insectivores). It must also be noted that there are more woodpeckers and bark gleaners in Guiana (the latter being passerines: woodcreepers and allies). There are also fewer foliage gleaners, but more birds specialized in the exploitation of accumulated dead or even decaying leaves retained locally in the foliage. Again, in Guiana, the ant-following guild appears somewhat richer: more species than in Gabon specialize in the exploitation of prey driven out by army ants. It can also be mentioned here that in Gabon, one hornbill *Tropicranus albocristatus* (Bucerotidae) associates regularly with the troops of monkeys, a behavior that has brought about a number of remarkable ecomorphological adaptations. Raptors of the genus *Harpagus* (Accipitridae) are also known in Guiana to follow troops of monkeys, but it is still unclear whether their association is as permanent and as close as that of *Tropicranus*.

If these results are compared with those of Terborgh, taking into account possible differences in guild definition, it becomes apparent that the same guilds are found in Guiana and in eastern Peru, leaving aside the aquatic birds and the frugivores/insectivores/nectarivores. This similarity would tend to indicate that the rainforests of the two study sites have similar structures. However, some guilds differ in their species richness. There are more numerous specialized raptors, fewer foliage gleaner insectivores, more salliers and aerial feeders, and also more species with a mixed diet in Peru than in Guiana. Also apparent is a higher proportion of frugivores *sensu lato* in Peru -36.7% (76/207 species) vs. 29.5% (67/227) $-$ while the proportions of partly and exclusive insectivores are almost the same -64.7% (133/207) vs. 61.2% (139/227).

A comparison between Peru and Gabon makes the differences observed between Gabon and Guiana even more apparent, the respective percentages rising to 36.7% vs. 16% (28/175 species) for frugivores, and to 64.7% vs. 73.7% (129/175) for insectivores.

It therefore seems that the two sites are hardly comparable and that their existing similarities can more readily be interpreted in terms of historical factors than ecological convergences. Such a remark applies particularly well to the comparison between Peru and Guiana. Nevertheless one cannot entirely rule out the possibility that these differences may, at least in part, be due to differences between authors in what constitutes a true forest species, or not $-$ or to slight differences among different forest types. A closer inspection of Terborgh's table makes me suspect that a number of species that I did not consider as true forest species, and therefore did not take into account in my calculations, were included in his list.

When Beehler's (1981) data are included to broaden the comparison, it appears that

1. Forest carrion eaters are present only in South America (Cathartidae).
2. Specialized raptors are particularly numerous at the Peruvian site.
3. Frugivores are more abundant in Papua New Guinea (P.N.G.) than in South America; this gives for P.N.G. a lower proportion (52.3%) of insectivores (partial and exclusive).
4. Terrestrial frugivores are even more numerous in P.N.G. than in South America. Beehler suggests that these birds have partly filled the niche of terrestrial mammals; however, this hypothesis is not supported by the comparison of the mammal faunas of Gabon and Guiana (see Dubost, 1987).
5. The P.N.G. bird community includes more nectarivores than those of Guiana and Gabon.
6. The number of bark-gleaning insectivores is much lower in Gabon and in P.N.G.
7. Woodpeckers, ant-following species, and birds specializing in the exploitation of lianas and accumulations of dead leaves in foliage are lacking in P.N.G. However, as pointed out by Jiro Kikkawa (in litt.), in P.N.G. forests other than those studied by Beehler there are bird species using lianas exclusively as a feeding substrate though "dead-leaves-caught-in-trees" specialists are absent.

Comparisons Between Tropical and Temperate Habitats

In an attempt to broaden the scope of this study I will now extend the comparison of community structure between the two tropical rainforests I studied personally by including a temperate "primeval" forest, namely, Bialowieza in Poland. This comparison is based on the recent paper of Tomialojc et al. (1984), and more particularly its appendix, which lists all the breeding species. From this list, 19 species were excluded as not truly forest species (e.g., *Ciconia nigra*, *Anas platyrhynchos*, *Milvus migrans*, and *Tringa ochropus*). This done I ended up with a list of 56 species characteristic of broad-leaved deciduous, evergreen coniferous, and mixed forests. They were allocated to the various guilds previously used (Table 4.4). This first task was not as simple as it might appear as temperate birds are not easily allocated to the same guilds as tropical birds; they appear to be less specialized and more flexible in their food choices and foraging behaviors. Thus tits (*Parus*) have been considered foliage gleaners in spite of the fact that at least some species search bark, although they are not true bark-gleaning specialists as are tree creepers (*Certhia*) and nuthatches (*Sitta*). Among woodpeckers, *Picoides major* has been included in the "frugivores/insectivores, arboreal gleaners" guild rather than among the "insectivores, woodpeckers" because it is sedentary and its diet includes many hard fruits when available (such as acorns and nuts). Being unacquainted with the local feeding habits of *Dryocopus martius*, I allocated this species to the woodpeckers guild, though it could just as well be considered a terrestrial insectivore because of its habit of opening large anthills. The jay *Garrulus glandarius* has been considered here as a frugivore/predator in spite of its omnivorous diet. Likewise the raven *Corvus corax* has been classified as a carrion eater though it is also an omnivore; it is the only species feeding on animal carcasses in this temperate bird community. Similar comments could be made for almost every other species on the Polish list.

I must also stress that, in a number of cases, it has been quite difficult to distinguish terrestrial from arboreal species. These two categories are not as clear-cut in temperate habitats as they are in tropical forests. We must therefore keep in mind that a comparison of temperate and tropical bird communities based on such broad-niche categories must be considered as only tentative.

Table 4.4 shows that when compared with the equatorial forests, the temperate forest of Bialowieza differs in the following ways:

1. It contains three or four times fewer bird species.
2. Several guilds are missing: nightjar, terrestrial and arboreal frugivores, nectarivores, aerial insectivores, ant followers, dead-leaf and vine gleaners, sallying arboreal frugivores/insectivores, and aquatic feeders. Representatives of these guilds account for 33.6% (76 species) of the Guiana list or 19.4% (34 species) of the Gabon list. Among the 24 guilds used to categorize all bird species at the three sites, none is peculiar to the temperate forest, while 10 are found in tropical habitats only.
3. It includes in absolute value but not as percentage of the whole avifauna a smaller number of diurnal and nocturnal raptors: a total of 7 species (12.6%), against 15 (8.6%) in Gabon, or 13 (5.7%) in Guiana.
4. It contains 71.3% (40 species) of fully or partially insectivorous birds, a proportion that is similar to that of Gabon.
5. It harbors only 12 species of strictly or partly frugivorous birds, nevertheless representing 21.4% of the avifauna—a higher proportion than in Gabon. These frugivores are essentially mast eaters: they feed on seeds, but their diet includes also such items as buds, seedlings, and young leaves of saplings.

6. It possesses 11 species of ground feeders, against 16 in Gabon or 22 in Guiana, but they represent 19.6% of the avifauna—a particularly high proportion.
7. The Polish list is remarkable for its high proportion of woodpeckers. The presence of these birds can likely be explained by the fact that timber is not removed and regeneration occurs continuously in this Polish forest reserve. There are many dead trees of various sizes left standing or decaying.
8. It has no forest species feeding entirely on aquatic organisms.
9. It has half of its avifauna made up of migrants (23 species out of 56!) and, among the resident species, only 15 are truly sedentary, spending the whole of the winter at Bialowieza; the others are erratic and leave the forest during the coldest winter episodes.

Discussion

I do not intend here to review again the current ideas on the causes of high bird species richness in the humid tropics as compared to temperate regions. I will nevertheless make some comments on Terborgh's (1980) paper, in which he discussed the respective roles played by guild niche breadth and by species packing on the guild dimension in the structuring of tropical bird communities. Considering a given guild, Terborgh tried to answer the following question: Can the greater bird species richness in tropical forests be due to the fact that guild members benefit from broader guild niches than in temperate forests, or is the high species richness due to a tighter species packing within guilds? Using the guilds already mentioned, he postulated that, in a given guild, species are ordered along a gradient of body weights in such a way that each species differs by a factor of roughly 2 from those preceding and following it. For every guild Terborgh measured its included niche using base 2 logarithms of the ratios of the heaviest to the lightest guild members. Species packing was defined as the quotient of the number of species in the guild minus 1; by guild-included niche. This allowed Terborgh to compute how many "extra" species in the guild could be allocated to a broad tropical niche as compared to temperate ones. Thus, for a given guild the product of tropical-included niche by the temperate species packing gives the number (minus 1) of species that would potentially make up the tropical guild, if species packing remained the same as in temperate regions. The difference between this calculated (plus 1) number and the number of species existing in the temperate guild accounts for the "extra tropical diversity" due to niche broadening. This resulting number is then subtracted from the total number of "extra" species to obtain the number of species whose presence can be interpreted in terms of a tighter species packing in tropical guilds.

Tables 4.7 and 4.8 illustrate the results to which Terborgh's method leads when Gabon and Poland, Guiana and Poland, and Guiana and Gabon are compared.

The existence in the tropics of guilds not present in temperate regions accounts for 28.1% of extra species in Gabon and 44.2% in Guiana. Terborgh found an intermediate value (34%) when comparing a central South Carolina forest with his Peruvian study site. The proportions of extra tropical species accountable by an enlargement of niche breadth are almost similar in Gabon (21.5%) and in Guiana (22.7%), though slightly higher than Terborgh's percentage for Peru (17%). If we now consider species packing

Table 4.7. Guild Structure of the Three Studied Bird Communities According to Weight Ranges of Component Species, and to Guild-Included Niche and Species Packing

Trophic/behavioral guild	Poland			Gabon			F. Guiana		
	W	I.n.	S.P.	W	I.n.	S.P.	W	I.n.	S.P.
Carrion	1150g	—	1	—	—	—	1450g	—	—
Raptor									
general	910–1400g	0.6	1.6	255–3650g	3.8	1.0	550–4500g	3.0	2.0
bird	200–950g	2.2	0.4	110–700g	2.7	0.7	280g	—	1
other specialist	765g	—	1	900g	—	1	200g	—	1
Owl	66–455g	2.8	0.3	130–1200g	3.2	1.6	70–500g	2.8	1.1
Nightjar	—	—	—	40–60g	0.6	1.7	40–55g	0.5	2.2
Mast eater									
terrestrial	140–530g	1.9	1.6	22–1600g	6.2	0.5	27–2985g	6.8	1.0
arboreal	13–34g	1.4	1.4	45–500g	3.5	1.1	25–1475g	5.9	2.7
Frugivore									
terrestrial	—	—	—	—	—	—	1070g	—	1
arboreal	—	—	—	45–1300g	4.8	2.1	10–1010g	6.6	3.7
Nectarivore	—	—	—	7g	—	1	2–18g	3.2	3.8
Insectivore									
terrestrial	9–315g	5.1	1.2	18–1100g	5.9	1.3	20–60g	1.6	7.6
woodpecker	22–300g	3.8	1.1	8–50g	2.6	1.5	13–250g	4.3	2.3
bark-gleaning	9–22g	1.3	0.8	35–45g	0.4	8.3	10–90g	3.2	4.7
foliage-gleaning	5–110g	4.4	3.6	7–260g	5.2	13.6	5–190g	5.2	7.6
sallying	10–16g	0.7	5.9	8–400g	5.6	3.5	6–145g	4.6	4.1
aerial	—	—	—	10–15g	0.6	1.7	14g	—	1
ant-following	—	—	—	35–60g	0.8	3.9	17–115g	2.8	2.2
dead leaf-gleaning	—	—	—	20g	—	1	9–31g	1.8	5.6
vine-gleaning	—	—	—	12–22g	0.9	3.4	8–20g	1.3	3.0
Frugivore/predator	175g	—	1	—	—	—	620g	—	1
Frugivore/insectivore									
arboreal, gleaning	55–175g	1.7	1.8	19–110g	2.5	2.0	14–310g	4.5	2.0
arboreal, sallying	—	—	—	275g	—	1	37–55g	0.6	5.2
Aquatic	—	—	—	35–915g	4.7	1.3	14–815g	5.9	0.8

Note: W = mean weights of the lightest and heaviest species in the guild; I.n. = included niche = $\log_2 (W_{max}/W_{min})$; S.P. = species packing = $(N-1)/I.n.$, with N = number of species in the guild.

Table 4.8. Species Richness Increases Accounted for by Presence of an Original Guild, by Included Niche Broadening, and by Species Packing Tightening, Respectively, as Revealed from Comparisons of Several Bird Communities

Trophic/behavioral guild	Gabon vs. Poland			F. Guiana vs. Poland			F. Guiana vs. Gabon		
	A	B	C	A	B	C	A	B	C
Carrion	(1)	–	–	–	–	–	1	–	–
Raptor									
general	–	3	–	–	4	1	–	–	2
bird	–	–	1	–	(1)	–	–	(2)	–
other specialist	–	–	–	–	–	–	–	–	(2)
Owl	2	–	4	–	–	2	–	–	–
Nightjar	2	–	–	2	–	–	–	–	–
Mast eater									
terrestrial	–	2	–	–	4	–	–	–	4
arboreal	–	–	–	–	6	8	–	3	9
Frugivore									
terrestrial	–	–	–	1	–	–	1	–	–
arboreal	12	–	–	26	–	–	–	4	10
Nectarivore	1	–	–	13	–	–	–	3	9

	A	B	C	A	B	C	A	B	C
Insectivore									
terrestrial	–	1	1	–	–	6	–	–	4
woodpecker	–	–	–	–	1	5	–	2	4
bark-gleaning	–	3	2	–	1	13	–	12	–
foliage-gleaning	–	16	52	–	3	21	–	–	(31)
sallying	2	–	–	1	15	–	–	(1)	–
aerial	4	–	–	7	–	–	–	–	(1)
ant-following	1	–	–	11	–	–	–	3	–
dead leaf-gleaning	4	–	–	5	–	–	–	–	10
vine-gleaning	(1)	–	–	–	–	–	–	1	–
Frugivore/predator	–	–	–	–	–	–	1	–	–
Frugivore/insectivore									
arboreal, gleaning	1	1	1	4	5	1	–	4	–
arboreal, sallying	7	–	–	6	–	–	–	1	2
Aquatic	–	–	–	–	–	–	–	–	(1)
	34	26	61	76	39	57	3	33	54
	(28.1%)	(21.5%)	(50.4%)	(44.2%)	(22.7%)	(33.1%)	(3.3%)	(37.6%)	(60.0%)

Note: A = number of extra species due to the presence in the richest avifauna of guilds not represented in the other; B = number of extra species accounted for by an enlargement of included niche; C = number of extra species attributable to an increased species packing. Figures in the parenthesis refer to number of extra species packing when the globally poorest avifauna possesses more species in the given guild than the richest one.

Table 4.9. Number of Guilds with a Higher Species Richness in One Country[a]

	A	B	C	D	E	Total A–E	F	Guild Total
F. Guiana vs. Gabon	3	6	–	5	8	22	2	24
F. Guiana vs. Poland	10	3	2	4	2	21	3	24
Gabon vs. Poland	11	3	–	3	3	20	4	24

[a] Because they are present there but not in the other country (A), or have a broader included niche (B), or more species are accounted for there by wider included niche than by tighter species packing (C), or the reverse (D), or have only a stronger species packing (E). The number of guilds without difference between the two compared countries is indicated in (F).

it appears that its tightening in the tropics would account for 50.4% of extra species in Gabon, but only for 33.1% in Guiana, though the increase in absolute species richness is almost the same for both countries: 57 species in Guiana, 61 in Gabon. Terborgh finds that 49% of the increase in species richness is due to a tighter species packing in Peru.

When comparing the Guiana and Gabon sites one notes that, although Guiana has more species than Gabon, only three of them (3.3%) belong to neotropical guilds missing at the Afrotropical site. Therefore 36.7% of Guianan extra species would be due to niche broadening and 60% to a tighter species packing.

If we rank the guilds according to the ways they differ between sites (Table 4.9), it appears that as many guilds differ in intertropical comparisons as in temperate vs. tropical comparisons. On the other hand, it appears that while tropical sites harbor more specific guilds than temperate sites, the differences in species richness are proportionally the same between Gabon and Guiana as between Gabon and Poland: 6 guilds with broader niches, and 13 with tighter species packing, against 3 and 6, respectively, in the second case, while the difference between Guiana and Poland is 5 vs. 6. It thus appears that the respective influences of the increase in numbers of specific guilds, of niche broadening, and of species-packing changes according to the comparisons made. The only emerging patterns are (1) the number of guilds in the tropics larger than in temperate regions, (2) the fact that less than a quarter of extra tropical species can be accounted for by niche breadth enlargements, and (3) the evidence that tightening of species packing varies greatly from one tropical site to another.

Following these general considerations, and still using Terborgh's method, more specific comments can be made if one compares the avifaunas in the various sites, with emphasis on their broad trophic categories such as frugivores, insectivores, and mixed feeders (Tables 4.10 and 4.11). Then it appears that whereas tropical sites harbor similar percentages of insectivorous birds and thus differ from the temperate site, no such difference is apparent for frugivores and mixed feeders. In this case, there exist as many differences between Gabon and Guiana as between Gabon and Poland. It can also be noted that every increase in species richness of insectivores is correlated with a tightening of species packing. It is also interesting to note that, according to Table 4.5, Guiana has 18 more species of frugivorous passerines and, conversely, 15 fewer nonpasserines only. This is a strong argument in favor of the importance of historical factors in the origin and structuring of avian communities.

Table 4.10. Weight Range, Included Niche, and Species Packing for Guilds of True Frugivores, True Insectivores, and Mixed Diet Species

	Poland				Gabon				F. Guiana			
	N	W	I.n.	S.P.	N	W	I.n.	S.P.	N	W	I.n.	S.P.
Frugivores	7	13–530g	5.3	1.1	21	22–1600g	6.2	3.2	52	10–2985g	8.2	6.2
Mixed	5	55–175g	1.7	2.4	7	19–275g	3.8	1.6	15	14–620g	5.5	2.6
Insectivores	36	5–315g	6.0	5.9	122	7–400g	5.8	20.7	125	5–250g	5.6	22.0

Note: For abbreviations see Table 4.7.

Table 4.11. Contribution (Number of Extra Species) of Increases of Included Niche or Species Packing to Increases of Species Richness

	Frugivores			Mixed			Insectivores		
Extra Species Due to	Gabon vs. Poland	F. Guiana vs. Poland	F. Guiana vs. Gabon	Gabon vs. Poland	F. Guiana vs. Poland	F. Guiana vs. Gabon	Gabon vs. Poland	F. Guiana vs. Poland	F. Guiana vs. Gabon
Included niche broadening	1	3	7	2	9	3	–	–	–
Species packing tightening	13	42	24	–	1	5	86	89	3
Total	14	45	31	2	10	8	86	89	3

An increase in species packing seems to be very important in explaining the increase in species richness of frugivores, but the broadening of included niches is also relevant; its role is probably much less important, although it is much more obvious when different tropical sites are compared than when comparisons are made between tropical and temperate sites.

Conversely, the increase in species richness of mixed feeders could well be essentially due to a broadening of their included niches, at least when the humid tropics are compared to the temperate zones; a comparison of Guiana and Gabon shows that both the tightening of species packing and a broadening of the included niches exert a similar influence on species richness.

If the frugivore and the frugivore/insectivore guilds are split into seed dispersers and seed predators (Tables 4.12 and 4.13), the two "subguilds" have very similar included niches in Guiana and Gabon, whereas seed dispersers in Poland have an included niche amounting to only one-third that of seed predators. Terborgh's method therefore leads me to conclude that a tightening of species packing plays the major role in species richness increase for seed predators, and also for seed dispersers when Gabon and Guiana are compared. On the other hand, the increase in the number of seed dispersers in the tropics, when compared to temperate regions, could be better accounted for by a broadening of their included niches. It is also worth noting that there are many more seed dispersers in Gabon than in Poland, and that they are distributed along a wider size range, while Guiana differs as much from Gabon as from Poland for these two categories of frugivores.

It would be tempting to compare in detail the various guilds and to proceed further in the analysis of the respective contributions of the broadening of included niche and of species packing to the intersite differences in species richness. However, Terborgh's method has its limitations that do not allow this.

(1) First of all, the choice of guild categories can be questioned. I have already mentioned the difficulties I encountered in applying tropical niche categories to temperate birds. This is due to the fact that our knowledge of the behavior and ecology of most bird species is still far from being satisfactory in the tropics, and even in some temperate habitats. It must also be stressed that in order to remain quite comparable between sites, guilds should be rigorously defined by using the same criteria. This implies of course a detailed knowledge of the ecology and behavior of the various species, but also that such a knowledge does really apply to the populations of the communities concerned. Data on foraging behavior and diet should relate to the population under study and not—as I have done here for Poland and (in part) for Guiana— drawn from the literature.

I could also elaborate on the concept of guild structure, but this would lead me too far from the present comparison (see Adams, 1985; Kikkawa, in press; Kikkawa and Williams, 1971; Wiens, 1983). I shall only say that guilds, as used here, can only give a rough picture of the manner in which species can be grouped together according to their similar ways of exploiting the same kind of food resources. A much better knowledge of the species would obviously allow a finer and more realistic guild partitioning that would make comparisons easier and more meaningful. Obviously, guilds like "insectivores, foliage gleaners" or "insectivores, salliers" group together species that differ both in ecology and behavior. A finer guild definition based on a better niche allocation of species would allow the distinction of species assemblages depending on mere phyletic relationships, environmental discontinuities, and interactions between species.

(2) Another important point relative to the use of guilds sensu Terborgh (1980) is that of the average live body weight of the various species. Our present knowledge of the tropical avifaunas allows us to use only crude figures for most species. In this study I had to use weights from the literature for Polish birds, some of the Guiana species, and even for a few of the Gabonese ones. Even when we have at our disposal figures from our own study sites, we still are far from being able to take into consideration individual, intrapopulational, and seasonal variations, and to compute realistic averages. This is an important drawback, especially for the smaller species of a guild, because the use of too crude estimates of body weight might have a strong incidence on the computation of included niches. The weight limits I used for establishing my various tropical guilds correspond to figures that have been at times rounded off, though they were always taken in observed ranges.

(3) The previous comments lead me to note that when Terborgh admits, following Hutchinson (1959), that inside a given guild species are broadly ordered by doubling their weight along a weight gradient, he suggests a constant size ratio between co-occurring congeneric or potentially competitive species. Too many ecologists have turned Hutchinson's hypothesis into a dogma but this view is now increasingly questioned (e.g., Horn and May, 1977; Strong et al., 1979; Roth, 1981; Simberloff and Boecklen, 1981; Wiens and Rottenberry, 1981; Simberloff, 1983; Erard, 1987). Accordingly, should we reject Terborgh's analysis as well as my own? My present personal experience of tropical avifaunas suggests that within the various guilds, species are not distributed along size or weight gradients according to a geometrical or arithmetical progression. However, pending a better knowledge of their diet and behavior, and in the absence of an adequate statistical analysis of weight distributions within objectively defined guilds (after live weights have been adequately sampled!), it seems impossible at the moment to decide whether weights of guild members follow some mathematical rule of proportionality or whether they are merely distributed at random.

As interactions between taxonomically, morphologically, and/or ecologically similar species have been observed (see, for instance, the case of the flycatchers (Muscicapidae) at the Gabon site; Erard, 1987 and in prep.), I think that Terborgh's viewpoint, though not rigorously testable, allows one at least to draw a crude yet useful picture of the various guild niche breadths; it probably also indicates the lowest limit of the contribution of included niche broadening, and the highest limit of the contribution of species packing tightening to the problem of the increased species richness of tropical avifaunas.

(4) It must also be noted that the concepts of guild, niche breadth, and species packing result from a body of theory that gives interspecific competition a major role in the structuring of ecological communities (see Vuilleumier, 1979). As a matter of fact, as I pointed out in the introduction, the idea that competition is always of paramount importance is now more and more questioned. Discussions concern not only the assumptions of the theory but also the methods used to analyze data (e.g., Wiens, 1983 and the various papers in *American Naturalist* 1983 and in Strong et al., 1984). Further-

Table 4.12. Weight Ranges, Included Niches, and Species Packings of Dispersers and Predators of Seeds

	Poland				Gabon				F. Guiana			
	N	W	I.n.	S.P.	N	W	I.n.	S.P.	N	W	I.n.	S.P.
Seed predators	7	13–530g	5.3	1.1	9	22–1600g	6.2	1.3	25	25–2985g	6.9	3.5
Seed dispersers	5	55–175g	1.7	2.4	19	19–1300g	6.1	3.0	42	10–1070g	6.7	6.1

Note: For abbreviations see Table 4.7.

Table 4.13. Contribution (Number of Extra Species) of Increases of Included Niche or Species Packing to Increases of Species Richness for Seed Predators and Dispersers

	Seed Predators			Seed Dispersers		
Extra Species Due to	Gabon vs. Poland	F. Guiana vs. Poland	F. Guiana vs. Gabon	Gabon vs. Poland	F. Guiana vs. Poland	F. Guiana vs. Gabon
Included niche broadening	1	2	1	10	12	2
Species packing tightening	1	16	15	4	25	21
Total	2	18	16	14	37	23

more such a theory postulates that the assemblages of species are always at, or close to, equilibrium. Moreover, the optimal structuring and composition of communities is supposed to follow the temporal variations of environmental conditions. However, as pointed out by Wiens (1984, p. 451), "natural communities should be viewed as being arrayed along a gradient of states ranging from nonequilibrium to equilibrium." Indeed, it has long been claimed that tropical forests bird communities were at equilibrium, although no long-term study has ever definitely documented such an assumption. Nevertheless, 20 years experience with the birds of Gabon (Brosset and Erard, in prep.) suggests that the rainforest bird communities we observed there did not change their structure much during these two decades. The changes we did notice were associated with the physiognomic changes of the vegetation due to human interferences. For example, the establishment of a large clearing for the buildings of the M'Passa field station had apparently increased the influence of tropical storms, resulting in a progressive transformation of a part of the high forest of the plateau into a lower, liana-rich forest full of gaps. In the untouched portions of the forest, however, the bird community has remained the same throughout the years (see, for instance, Erard, 1987 for a 5-year intensive study of flycatchers). The various bird species of the Gabon rainforest are remarkably sedentary and apparently do not disperse very far (this point must be more precisely documented, however), an observation that suggests that the overall pattern is one of relative stability of the community. Conversely, although part of the forest community members are sedentary in Guiana as in Gabon, many frugivores or mixed feeders appear to be much more erratic and range rather far. A long-term perspective is definitely needed to obtain a better view of the stability of tropical rainforest bird communities. Furthermore, although relatively stable, the community structure does vary slightly at least seasonally.

Moreover, while at the North-East Gabon scale, the forest avifauna (excluding birds of second growth) appears to be relatively homogeneous, its species richness and composition remaining much the same from site to site (though variations can occur in species abundances), the situation is different in Guiana. Here, important intersite differences do occur (personal observation and Thiollay, 1986). These considerations introduce an important element of uncertainty in our search for processes explaining observed differences in species richness. This leads to another remark.

(5) In the comparison made so far, the most debatable point seems to be the definition of the habitats themselves. The list of birds of the Bialowieza forest includes in fact taxa living in a mosaic of broad-leaved, evergreen, and mixed forests. This means that bird diversity is most probably of a γ (landscape) type rather than strictly an α (within-habitat) type diversity. The resulting overall species richness is that of a juxtaposition of different habitats, not that of a single one. This might account in part for the higher species richness of Bialowieza as compared to Terborgh's South Carolina forest. Moreover, I do not have for Poland a list of bird species resident on a single 2-km² forest plot, though the difference might be slight as shown by the different lists drawn for the various areas sampled by Tomialojc et al. (1984).

The mosaic problem exists in the tropics too. Although I took into consideration in Gabon as well as in Guiana only natural "primary" forest, neglecting species associated with forest edges and secondary successions, the Gabonese and Guianese sites remain nevertheless quite different. Compared to that of North-East Gabon, the Guiana forest studied looks more like a mosaic of different forest types (upland, unflooded, primary forest; riparian or swampy successions; palm stands; liana-rich lowland forest with low canopy; and other types). All of these forest types form patches of a few hectares each, and together they participate in larger tridimensional forest mosaic. Moreover, the canopy is generally less dense, more open, in Guiana than in Gabon, a characteristic that added to the presence of numerous palm trees of various species, makes the understorey and the forest floor relatively better lighted. This means that the Guiana forest studied appears to be structurally more diverse than that of North-East Gabon. In the absence of detailed descriptions of the habitats studied in such forests, based on quantitative analyses of vegetation characteristics, it is therefore almost impossible to infer any process from patterns emerging from species richness differences at the landscape scale.

Conclusion

Following this discussion, it might be tempting to doubt the usefulness of intersite comparisons based on such broad and superficial approaches of ecosystems as complex as humid tropical forests. However, the comparisons between Gabon and Guiana were made by the same observer who worked in the same way at both sites and used the same methods for collecting and analyzing the data. This could contribute to circumventing, or at least strongly reducing, some of the sampling biases, and enhance the value of the patterns detected.

I will not elaborate any further on the conclusions of my previous paper (Erard, 1986) in which I emphasized the differences in taxonomic structure of the rainforest bird communities of Gabon and Guiana, the dissimilarities in the occupation of habitat between the two sites, and the greater number of frugivorous species in Guiana.

The more elaborate analysis carried out in this chapter illustrates the fact that intersite differences between communities are not only large between tropical and temperate sites, but also exist between African and South American ones. Although the allocation of species to distinct guilds ends up by producing similar sets of categories in Guiana and in Gabon (whose kinds and numbers are, of course, quite different from those in Poland), the number of species belonging to these functional units varies greatly from site to site. If the present findings reflect the natural situation they not only indicate that a limited number of adaptive zones is available to forest birds but also that the relative importance of each of these adaptive zones varies from continent to continent, and perhaps even from one forest to the other in a given biome (see also Blondel et al., 1985 for Mediterranean birds).

A number of remarks can also be made on the diets and weight classes of the bird species belonging to the communities under study. It has long been recognized that frugivorous birds are more numerous in tropical than in temperate forests. However, there are important differences between tropical sites; sets of large fruit-eating species exist in Guiana that are lacking in Gabon. Furthermore, in the latter country, frugivory is essentially the concern of nonpasserine birds, and there is a balance between seed dispersers and seed predators, whereas in Guiana there are almost as many frugivorous passerines as nonpasserines, the former being almost all seed dispersers and the latter seed predators. This would suggest an important difference in the quality and quantity of fruit production between two sites. The fact that 32 entirely or partially frugivorous passerine species live at the Guiana site, against only 7 in Gabon, might suggest some intersite difference in the number of the small fruits available for birds (see the syndromes defined by Gautier-Hion et al., 1985a); this remains to be studied. These differences in the structure of frugivore guilds will probably be better understood when the interactions taking place between various frugivores, and those between fruiting plants and frugivorous vertebrates, will be better known. Such relationships can be interpreted in terms of a diffuse coevolution based on mutualistic and/or antipredator interactions (see, for example, Janzen, 1983; Herrera, 1985; Erard and Sabatier, in press; Wheelwright, in press).

Very large predatory forest bird species exist in the four tropical forest habitats studied but they are lacking at Bialowieza. They are eagles hunting monkeys and ground mammals, such as small duikers, in Africa and large rodents in South America. The presence of fishing owls in Gabon and their absence in the other tropical sites studied is worthy of note; is their niche filled by members of other animal groups?

Ground insectivores are small species (they are all passerines) in Guiana, contrary to what happens in Gabon and, to a lesser extent, in Poland. This is in contrast to the fact that the ground frugivores (seed dispersers and seed predators) reach larger sizes in Guiana than in the other countries, and are represented by nonpasserines. This suggests that the availability of different kinds of fruits on both continents has influenced the evolutionary history of the bird communities. Apparently, nonpasserine birds were able to specialize on fruit in Guiana, whereas in Gabon these birds became in almost equal proportion either insectivores or frugivores. Again among insectivores, the small number of woodpecker species, and especially the lack of larger species, must also be noted in Gabon. This can be explained by the past history of the Picidae and, more particularly, of the Campephilinae (see Short, 1982).

Salliers range over a wider weight range in tropical than in temperate forests. The largest species are found in Gabon (where they are insectivores) and in Guiana (where they are frugivores/insectivores). Here again, we note the tendency for the true and mixed frugivores to reach larger sizes in Guiana than in Gabon, whereas the reverse is true for insectivores (see Table 4.10).

Obviously, historical factors played a paramount role in the origin and structuring of all the bird communities discussed in this paper. An important difference in the organization of these communities is to be traced into the different evolution of frugivory on both continents. Many of the differences between African and neotropical bird communities might very likely be attributable to factors such as the past geographic distribution of habitats and their degree of fragmentation during the Pleistocene-Holocene, the long-term and short-term effects of climatic changes, the distribution and importance of dispersal barriers, and the influence of past human disturbances on the habitats and their faunas. All these factors remain to be adequately documented.

Although striking ecological, behavioral, and morphological convergences do exist between the forest avifaunas of tropical continents, especially among salliers (i.e., Meropidae/Galbulidae, Muscicapidae/Tyrannidae; see Fry, 1970; Keast, 1972), these convergences concern only species pairs, or small groups of species, and never entire guilds (see also Blondel et al., 1985; and Dorst and Vuilleumier, 1986). However, it would be important to undertake comprehensive studies of the morphological, physiological, and dispersal potential of the different species; this would allow a better understanding of the constraints habitat exerts on species design features and would help to determine the role played by species morphology in the organization of communities. (See, for example, Leisler, 1977, 1980; Leisler and Winkler, 1985; and the case of the Australian mangrove bat communities studied by McKenzie and Rolfe, 1986.)

The role predators play in structuring bird communities (see Connell, 1971, 1975) needs also to be better investigated. What are the major predators in each habitat? On what kind of species do they prey? When and how? What is their actual impact on specific populations and on community organization? For the time being there is no way of answering such questions. It is likely that, as suspected by Holt (1984), Blondel (1986), and Schoener (1986), among others, the mechanisms of predation and competition are not exclusive of each other, but are interactive and complementary components of the system that determines interspecific coactions.

Acknowledgments. I am particularly grateful to Professor F. Bourlière, Dr. A. Brosset, Professor J. Kikkawa, and Dr. F. Vuilleumier for their useful comments on an earlier draft of this chapter.

References

Adams, J. 1985. The definition and interpretation of guild structure in ecological communities. J. Anim. Ecol. 54, 43–59

Amadon, D. 1973. Birds of the Congo and Amazon forests: A comparison. In: Meggers, B.J., et al. (eds.), Tropical Forest Ecosystems in Africa and South America: A Comparative Review, Smithsonian Institution Press, Washington, D.C., pp. 267–277

Beehler, B. 1981. Ecological structuring of forest bird communities in New Guinea. Monographiae Biologicae 42, 837–861

Blondel, J. 1986. Biogéographie Évolutive, Masson, Paris

Blondel, J., Ferry, C., and Frochot, B. 1973. Avifaune et végétation, essai d'analyse de la diversité. Alauda 41, 63–84

Blondel, J., Vuilleumier, F., Marcus, L.F., and Terouanne, E. 1984. Is there ecomorphological convergence among Mediterranean bird communities of Chile, California, and France? Evol. Biol. 18, 141–213

Bourlière, F. 1983. Animal species diversity in tropical forests. In: Golley, F.B. (ed.), Tropical Rain Forest Ecosystems. A. Structure and Function, Elsevier, Amsterdam, pp. 77–91

Bourlière, F. 1984. Species richness in tropical forest vertebrates. Biol. Int., n° sp. 6, 49–60

Brackenbury, J.H. 1982. The structural basis of voice production and its relationship to sound characteristics. In: Kroodsma, D.E., and Miller, E.H. (eds.), Acoustic Communication in Birds, Vol. 1, Academic Press, New York, pp. 53–73

Brosset, A., and Erard, C. 1986. Les oiseaux des régions forestières du Nord-Est du Gabon. Vol. 1. Ecologie et comportement des espèces. Rev. Ecol. (Terre Vie), Suppl. 3, 1–297

Caballé, G. 1978. Essai sur la géographie forestière du Gabon. Adansonia 17, 425–400

Caballé, G. 1984. Essai sur la dynamique des peuplements de lianes ligneuses d'une forêt du Nord-Est du Gabon. Rev. Ecol. (Terre Vie) 39, 3–36

Connell, J.H. 1971. On the role of natural enemies in preventing competitive exclusion in some marine animals and in rain forest trees. In: de Boer, P.J., and Gradwell, G.R. (eds.), Dynamics of Populations, PUDOC, Wageningen, pp. 298–312

Connell, J.H. 1975. Some mechanisms producing structure in natural communities: A model and evidence from field experiments. In: Cody, M.L., and Diamond, J.M. (eds.), Ecology and Evolution of Communities, Harvard Univ. Press, Cambridge, MA, pp. 460–490

Connell, J.H., and Orias, E. 1964. The ecological regulation of species diversity. Am. Nat. 98, 399–414

Connor, E.F., and Simberloff, D. 1984. Neutral models of species' co-occurrence patterns. In: Strong, D.R., et al. (eds.), Ecological Communities: Conceptual Issues and the Evidence, Princeton Univ. Press, Princeton, NJ, pp. 316–331

Cracraft, J. 1973. Continental drift, paleoclimatology, and the evolution and biogeography of birds. J. Zool. Lond. 169, 455–545

Cruiziat, P. 1966. Note sur le microclimat de la strate inférieure de la forêt équatoriale comparé à celui d'une clairière. Biol. Gabonica 2, 361–402

Diamond, J.M. 1973. Distributional ecology of New Guinea birds. Science 179, 759–769

Diamond, J.M. 1978. Niche shifts and the rediscovery of interspecific competition. Am. Scient. 66, 322–331

Dobzhansky, T. 1950. Evolution in the tropics. Am. Scient. 38, 209–221

Dorst, J. 1973. Distribution des familles dominantes au sein de l'avifaune d'Amérique du Sud, d'Afrique et d'Australie, et ses relations avec l'évolution dans ces masses continentales. C.R. Acad. Sc. Paris 277(D), 1773–1777

Dorst, J. 1974. Hypothèses sur les causes de la diversification et de la richesse spécifique de l'avifaune néotropicale. C.R. Acad. Sc. Paris 278(D), 2535–2540

Dorst, J., and Vuilleumier, F. 1986. Convergences in bird communities at high altitudes in the tropics (especially the Andes and Africa) and at temperate latitudes (Tibet). In: Vuilleumier, F., and Monasterio, M. (eds.), High Altitude Tropical Biogeography, Oxford University Press and American Museum of Natural History, New York, pp. 120–149

Dubost, G. 1987. Une analyse écologique de deux faunes de Mammifères tropicaux. Mammalia 51, 415–436

Erard, C. 1986. Richesse spécifique de deux peuplements d'oiseaux forestiers équatoriaux: Une comparaison Gabon-Guyane. Mém. Mus. Nat. Hist. Nat., A, Zool. 132, 53–66

Erard, C. 1987. Ecologie et comportement des Gobemouches (Aves: Muscicapinae, Platysteirinae, Monarchinae) du Nord-Est du Gabon. Vol. 1. Morphologie des espèces et organisation du peuplement, Mém. Mus. Nat. Hist. Nat., A, Zool. 138, 1–256

Erard, C., and Sabatier, D. In press. Rôle des oiseaux frugivores terrestres dans la dynamique forestière en Guyane française. Proc. Int. Orn. Congr. 19

Florence, J. 1981. Chablis et sylvigénèse dans une forêt dense humide sempervirente du Gabon. Thèse de 3e Cycle. Université Louis Pasteur, Strasbourg

Fry, C.H. 1970. Convergence between jacamars and bee-eaters. Ibis 112, 257–259

Gasc, J.-P. 1986. Le peuplement herpétologique d'Astrocaryum paramaca (Arecacées), un palmier important dans la structure de la forêt en Guyane francaise. Mém. Mus. Nat. Hist. Nat., A, Zool. 132, 97–107

Gautier-Hion, A., Duplantier, J.-M., Quris, R., Feer, F., Sourd, C., Decoux, J.-P., Dubost, G., Emmons, L., Erard, C., Hecketsweiler, P., Moungazi, A., Roussilhon, C., and Thiollay, J.-M. 1985. Fruit character as a basis of fruit choice and seed dispersal in a tropical forest vertebrate community. Oecologia (Berlin) 65, 324–337

Gautier-Hion, A., Duplantier, J.-M., Emmons, L., Feer, F., Hecketsweiler, P., Moungazi, A., Quris, R., and Sourd, C. 1985. Coadaptation entre rythmes de fructification et frugivorie en forêt tropicale humide du Gabon: mythe ou réalité? Rev. Ecol. (Terre Vie) 40, 405–434

Gilpin, M.E., and Diamond, J.M. 1984. Are serious co-occurence on islands non-random, and are null hypotheses useful in community ecology? In: Strong, D.R., et al. (eds.), Ecological Communities: Conceptual Issues and the Evidence, Princeton Univ. Press, Princeton, NJ, pp. 297–315

Grant, P., and Schluter, D. 1984. Interspecific competition inferred from patterns of guild structure. In: Strong, D.R., et al. (eds.), Ecological Communities: conceptual Issues and the Evidence, Princeton Univ. Press, Princeton, NJ, pp. 201–233

Guillotin, M. 1981. Données écologiques sur les petits rongeurs forestiers terrestres de Guyane française. Thèse de 3e Cycle, Montpellier

Haffer, J. 1969. Speciation in Amazonian forest birds. Science 1965, 131–137

Haffer, J. 1974. Avian speciation in tropical South America. Nuttall Orn. Cl. Publ. 14, 1–390

Hallé, F., Oldeman, R.A.A., and Tomlinson, P.P. 1978. Tropical Trees and Forests: An Architectural Analysis. Springer-Verlag, Berlin

Hamel, P.J. 1980. Avifauna of the Kifu and Mabira forests, Uganda. Proc. Pan-African Orn. Congr 4, 135–145

Herrera, C.M. 1985. Determinant of plant-animal coevolution: The case of mutualistic dispersal of seeds by vertebrates. Oikos 44, 132–141

Hespenheide, H.A. 1971. Food preference and the extent of overlap in some insectivorous birds, with special reference to the Tyrannidae. Ibis 113, 59–72

Hladik, A. 1978. Phenology of leaf production in rain forest of Gabon: Distribution and composition of food for folivores. In: Montgomery, G.G. (ed.), The Ecology of Arboreal Folivores, Smithsonian Institution Press, Washington, D.C., pp. 51–71

Hladik, A. 1986. Données comparatives sur la richesse spécifique et les structures des peuplements des forêts tropicales d'Afrique et d'Amérique. Mém Mus. Nat. Hist. Nat., A, Zool. 132, 9–17

Holt, R.D. 1984. Spatial heterogeneity, indirect interactions and the coexistence of prey species. Am. Nat. 124, 377–406

Horn, H.S., and May, R.M. 1977. Limits to similarity among coexisting competitors. Nature 270, 660–661

Hutchinson, G.E. 1959. Homage to Santa Rosalia or Why are there so many kinds of animals? Am. Nat. 93, 145–159

Janzen, D.H. 1983. Dispersal of seeds by vertebrate guts. In: Futuyma, D.J., and Slatkin, M. (eds.), Coevolution, Sinauer Associates Inc., Sunderland, pp. 232–262

Karr, J. 1971. Structure of avian communities in selected Panama and Illinois habitats. Ecol. Monogr. 41, 207–233

Karr, J. 1975. Production, energy pathways, and community diversity in forest birds. In: Golley, F.B., and Medina, E. (eds.), Tropical Ecological Systems: Trends in Terrestrial and Aquatic Research. Springer-Verlag, New York, pp. 161–176

Karr, J. 1976a. Within and between habitat avian diversity in African and Neotropical lowlands habitats. Ecol. Monogr. 46, 457–481

Karr, J. 1976b. Seasonality, resource availability, and community diversity in tropical bird communities. Am. Nat. 110, 973–994

Karr, J. 1980. Geographical variation in the avifaunas of tropical forest undergrowth. Auk 97, 283–298

Karr, J., and James, F.C. 1975. Eco-morphological configurations and convergent evolution in species and communities. In: Cody, M.L., and Diamond, J.M. (eds.), Ecology and Evolution of Communities. Harvard Univ. Press, Cambridge, MA, pp. 258–291

Keast, A. 1972. Ecological opportunities and dominant families, as illustrated by the neotropical Tyrannidae (Aves). Evol. Biol. 5, 229–277

Kikkawa, J. 1982. Ecological association of birds and vegetation structure in wet tropical forests of Australia. Aust. J. Ecol. 7, 325–345

Kikkawa, J. In press. Bird communities of rainforests. Proc. Int. Orn. Congr. 19

Kikkawa, J., and Williams, W.T. 1971. Altitudinal distribution of land birds in New Guinea. Search (Sydney) 2, 64–65

Klopfer, P.H., and MacArthur, R.M. 1960. Niche size and faunal diversity. Am. Nat. 94, 293–300

Klopfer, P.H., and MacArthur, R.M. 1961. On the causes of tropical species diversity: niche overlap. Am. Nat. 95, 223–226

Leisler, B. 1977. Die ökologische Bedeutung der Lokomotion mitteleuropäischer Schwirle (Locustella). Egretta 20, 1–25

Leisler, B. 1980. Morphological aspects of ecological specialization in bird genera. Ökol. Vögel (Ecol. Birds) 2, 199–220

Leisler, B., and Winkler, H. 1985. Ecomorphology. Current Ornithol. 2, 155–186

Levins, R. 1968. Evolution in changing environments. Princeton Univ. Press, Princeton, NJ

Lovejoy, T.E. 1974. Bird diversity and abundance in Amazon forest communities. Living Birds 13, 127–191

MacArthur, R.M. 1958. Population ecology of some warblers of northeastern coniferous forest. Ecology 39, 599–619

MacArthur, R.M. 1969. Patterns of communities in the tropics. Biol. J. Linn. Soc. 1, 19–30

MacArthur, R.M. 1972. Geographical Ecology: Patterns in the Distribution of Species, Harper and Row, New York

MacArthur, R.M., and Levins, R. 1967. The limiting similarity, convergence, and divergence of coexisting species. Am. Nat. 101, 377–385

MacArthur, R.M., and MacArthur, J.W. 1961. On bird species diversity. Ecology 42, 594–598

MacArthur, R.M., MacArthur, J.W., and Preer, J. 1962. On bird species diversity. II. Prediction of bird censuses from habitat measurements. Am. Nat. 96, 167–174

MacArthur, R.M., Recher, H., and Cody, M. 1966. On the relation between habitat selection and species diversity. Am. Nat. 100, 319–332

Maury-Lechon, G., and Poncy, O. 1986. Dynamique forestière sur 6 hectares de forêt dense humide de Guyane française, à partir de quelques espèces de forêt primaire et de cicatrisation. Mém. Mus. nat. Hist. nat., A, Zool. 132, 211–242

McKenzie, N.L., and Rolfe, J.K. 1986. Structure of bat guilds in the Kimberley mangroves, Australia. J. Anim. Ecol. 55, 401–420

Moreau, R.E. 1966. The Bird Faunas of Africa and Its Islands, Academic Press, New York

Munn, C.A. 1985. Permanent canopy and understory flocks in Amazonia: Species composition and population density. Ornith. Monogr. 36, 683–712

Munn, C.A., and Terborgh, J.W. 1979. Multi-species territoriality in neotropical foraging flocks. Condor 81, 338–347

Oldeman, R.A.A. 1974. Ecotopes des arbres et gradients écologiques verticaux en forêt guyanaise. Terre Vie 28, 487–520

Orians, G.H. 1969. The number of bird species in some tropical forests. Ecology 50, 783–801

Paine, R.T. 1966. Food web complexity and species diversity. Am. Nat. 100, 65–75

Pearson, D.L. 1975. The relation of foliage complexity to ecological diversity of three Amazonian bird communities. Condor 77, 453–466

Pearson, D.L. 1977. A pantropical comparison of bird community structure on six lowland forest sites. Condor 79, 232–244

Pianka, E.R. 1966. Latitudinal gradients in species diversity: a review of concepts. Am. Nat. 100, 33–46

Powell, G.V.N. 1980. Mixed species flocking as a strategy for neotropical residents. Proc. Int. Orn. Congr. 17, 813–819

Powell, G.V.N. 1985. Sociobiology and adaptive significance of interspecific foraging flocks in the neotropics. Ornith. Monogr. 36, 713–732

Recher, H.F. 1969. Bird species diversity and habitat diversity in Australia and North America. Am. Nat. 103, 75–80

Root, R.B. 1967. The niche exploitation pattern of the Blue-gray Gnatcatcher. Ecol. Monogr. 37, 317–350

Roth, R.R. 1976. Spatial heterogeneity and bird species diversity. Ecology 57, 773–782

Roth, V.L. 1981. Constancy in the size ratios of sympatric species. Am. Nat. 118, 394–404

Sabatier, D. 1982. Périodicité de la fructification en forêt guyanaise. Bull. EC. ER. EX. 6, 149–164

Sabatier, D. 1983. Frutification et dissémination en forêt guyanaise. L'exemple de quelques espèces ligneuses. Thèse 3e Cycle, Montpellier

Sabatier, D. 1985. Saisonnalité et déterminisme du pic de fructification en forêt guyanaise. Rev. Ecol. (Terre Vie) 40, 289–320

Sabatier, D., and Puig, H. 1986. Phénologie et saisonnalité de la floraison et de la fructification en forêt dense guyanaise. Mém. Mus. nat. Hist. nat., A, Zool. 132, 173–184

Sanders, H.L. 1968. Marine benthic diversity: A comparative study. Am. Nat. 102, 243–282

122   C. Erard

Schoener, T.W. 1971. Theory of feeding strategies. Ann. Rev. Ecol. Syst. 2, 369–404
Schoener, T.W. 1983. Field experiments on interspecific competition. Am. Nat. 122, 240–285
Schoener, T.W. 1986. Resource partitioning. In: Kikkawa, J., and Anderson, D.J. (eds.), Community Ecology: Pattern and Process, Blackwell, Oxford, pp. 91–126
Short, L.L. 1980. Chaco woodland birds of South America: Some African comparisons. Proc. Pan-African Orn. Congr. 4, 147–158
Short, L.L. 1982. Woodpeckers of the World. Delaware Museum of Natural History, Greenville
Simberloff, D. 1983. Sizes of coexisting species. In: Futuyma, D.J., and Slatkin, M. (eds.), Coevolution. Sinauer Associates, Sunderland, pp. 404–430
Simberloff, D. 1984. Properties of coexisting bird species in two archipelagoes. In: Strong, D.R., et al. (eds.), Ecological Communities: Conceptual Issues and the Evidence, Princeton Univ. Press, Princeton, NJ, pp. 234–253
Simberloff, D., and Boecklen, W. 1981. Santa Rosalia reconsidered: Size ratios and competition. Evolution 35, 1206–1228
Smith, S.M. 1978. The "underworld" in a territorial sparrow: Adaptive strategy for floaters. Am. Nat. 112, 571–582
Strong, D.R., Simberloff, D., Abele, L.G., and Thistle, A.B. (eds.) 1984. Ecological Communities: Conceptual Issues and the Evidence, Princeton Univ. Press, Princeton, NJ
Strong, D.R., Szyska, L.A., and Simberloff, D.S. 1979. Test of community-wide character displacement against null hypotheses. Evolution 33, 897–913
Terborgh, J. 1977. Bird species diversity on an Andean elevational gradient. Ecology 58, 1007–1019
Terborgh, J. 1980. Causes of tropical species diversity. Proc. Int. Orn. Congr. 17, 955–961
Terborgh, J. 1985. Habitat selection in Amazonian birds. In: Cody, M.L. (ed.), Habitat Selection in Birds, Academic Press, New York, pp. 311–338
Terborgh, J., and Robinson, S. 1986. Guilds and their utility in ecology. In: Kikkawa, J., and Anderson, D.J. (eds.), Community Ecology: Pattern and Process, Blackwell, Oxford, pp. 65–90
Thiollay, J.-M 1986. Structure comparée du peuplement avien dans trois sites de forêt primaire en Guyane. Rev. Ecol. (Terre Vie) 41, 59–105
Tomialojc, L., Wesolowski, T., and Walankiewicz, W. 1984. Breeding bird community of a primaeval temperate forest (Bialowieza National Park, Poland). Acta Ornithologica 20, 241–310
Vuilleumier, F. 1969. Biotic diversity and environmental stability. Science 166, 210–211
Vuilleumier, F. 1970. Speciation in South American birds: A progress report. Act. IV Congr. Latin. Zool. 1, 239–255
Vuilleumier, F. 1972. Bird species diversity in Patagonia (temperate South America). Am. Nat. 106, 266–271
Vuilleumier, F. 1978. Remarques sur l'échantillonnage d'une riche avifaune de l'ouest de l'Ecuador. L'Oiseau et R.F.O. 48, 21–36
Vuilleumier, F. 1979. La niche de certains modélisateurs: Paramètres d'un monde réel ou d'un univers fictif ? Terre Vie 33, 375–423
Vuilleumier, F. 1980. Ecological aspects of speciation in birds, with special reference to South American birds. In: Reig, O.A. (ed.), Ecology and Genetics of Animal Speciation, Equinoccio, Univ. Simon Bolivar, Caracas, Venezuela, pp. 101–148
Wheelwright, N.T., In press. Tropical fruit-eating birds and their food plants: Evidence for coevolution? Proc. Int. Orn. Congr. 19
Wiens, J.A. 1977. On competition and variable environments. Am. Scient. 65, 590–597
Wiens, J.A. 1983. Avian community ecology: An iconoclastic view. In: Brush, A.H., and Clark, G.A. (eds.), Perspective in Ornithology, Cambridge Univ. Press, Cambridge, UK, pp. 355–403
Wiens, J.A. 1984. On understanding a nonequilibrium world: Myth and reality in community patterns and processes. In: Strong, D.R., and Simberloff, D. (eds.), Ecological Communities: Conceptual Issues and the Evidence, Princeton Univ. Press, Princeton, NJ, pp. 439–457
Wiens, J.A., and Rotenberry, J.T. 1981. Habitat associations and community structure of birds in shrubsteppe environments. Ecol. Monogr. 51, 21–41

5. Bird Species Diversity of Lowland Tropical Rainforests of New Guinea and Northern Australia

Peter V. Driscoll and Jiro Kikkawa

Introduction

Much of the rainforest on humid tropical lowlands has been cleared and what remains is under threat today in many parts of the tropics. The Australasian rainforests are no exception. In tropical Australia between Townsville and Cooktown 56.9% of the rainforest on the lowland plains has been destroyed since European settlement, and clearing is still continuing (Winter et al., 1987). In New Guinea, although 50% of forest occurs on the lowlands (McIntosh, 1974) there is little hope of retaining lowland rainforests in their pristine state if reservation is not realized and forest management not improved (White, 1971; Kwapena, 1985). Because the lowland fauna is the richest in the Australasian rainforest, it is important to document this fauna in terms of how it is affected by human disturbances to the forest.

In this chapter we examine the rainforest bird fauna of lowland New Guinea and in northeastern Australia to characterize their taxonomic composition, species richness and diversity, and ecological attributes. This provides a basis for comparison with the bird faunas of other tropical regions.

As shown in Figure 5.1, rainforest vegetation is much more extensive in New Guinea than in northern Australia. Also, the New Guinea rainforest fauna with a high degree of regional endemism is much richer than its Australian counterpart. In tropical Australia endemism of rainforest bird fauna is restricted to the tablelands. All rainforest species of Cape York Peninsula having distributions restricted to the north of Cooktown are conspecific with New Guinea forms (Kikkawa, 1976). Thus in comparing the

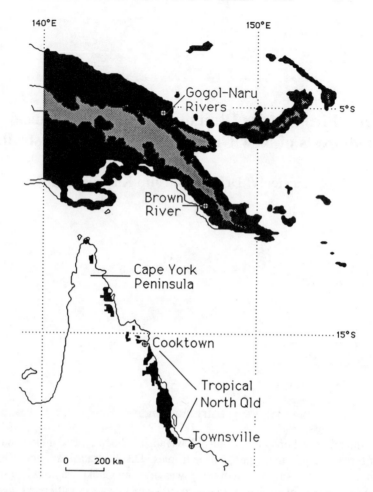

Figure 5.1. A map of Papua New Guinea and north Queensland showing the distribution of rain-forest. Black: lowland and tableland closed forests; stippled: montane forests; white: sclerophyll vegetation.

taxonomic composition of rainforest bird fauna between the New Guinea lowlands and tropical Australia we examine the faunas of two sites in New Guinea, repre-senting the ornithological subregions of Southeast and Sepik–Ramu (Beehler et al. 1986), and the entire rainforest fauna of northeastern Australia where three distinct faunal subprovinces may be recognized on the basis of distributional patterns (Kikkawa et al., 1981).

The pattern of niche occupation by rainforest birds is compared between subregions as in the taxonomic comparison, whereas local species richness and between-habitat diversity are analyzed within the Gogol and Naru Rivers area in New Guinea and between 4-ha plots in Cape York Peninsula and tropical north Queensland. Species overlap between habitats is obtained from comparisons of complete species lists for different sites within each subregion whereas between-habitat diversity is calculated from mist net data (Gogol–Naru and Cape York Peninsula plots) or dawn census data (tropical north Queensland plots). The plots represent areas of intact forests, either uniform or mosaic, and areas of impact caused by logging or other human activities, so that comparisons will at least indicate the degree of deviation in the faunal assemblage in areas of impact.

Taxonomic Composition

Table 5.1 gives taxonomic compositions of bird fauna by family for the Gogol–Naru area (5°15'S, 145°30'E) and Brown River (9°16'S, 147°05'E) in lowland New Guinea, and for the rainforest habitats of Cape York Peninsula between Cape York (10°41'S) and Cooktown (15°28'S), and tropical humid lowlands and tablelands of north Queensland between Cooktown and Townsville (19°16'S).

There are about 320 species of land birds recorded from the lowlands of New Guinea (Rand and Gilliard, 1967) and, of these, 147 species have been recorded from the Gogol–Naru area of the Sepik–Ramu subregion (Driscoll, 1985) and 184 species from a 2.5-ha plot at Brown River in the Southeast subregion (Bell, 1982a).

Cape York Peninsula north of Cooktown (about 120,000 km²) contains 260,260 ha of rainforest in small patches (Winter et al., 1987; Figure 5.1). Here, almost all (91%) of strictly rainforest species (69 species) are shared with New Guinea (Kikkawa, 1976). In addition to this attenuated New Guinea fauna, the rainforest patches of Cape York Peninsula receive birds of primarily sclerophyll habitats, which are usually absent from extensive rainforest habitats. The humid tropical regions of Queensland harbor the most extensive rainforest (Figure 5.1) in Australia, with 110,040 ha in area still remaining above 300 m in altitude and 269,800 ha on lowlands (Winter et al., 1987). The typical lowland fauna occurs on coastal plains (0–80 m in altitude) where nonrainforest species are sometimes associated with rainforest birds in natural mosaic habitats containing sclerophyll elements or in regrowth after logging. Tableland rainforests above 300 m, on the other hand, are usually well defined and exclude nonrainforest birds. Thus species richness of rainforest birds, when considered for each subregion as a whole, is greatest in Cape York Peninsula with 123 species, followed by humid tropical lowlands with 83 species, and least in tableland rainforests with 72 species.

The endemic families of the Australasian region include Casuariidae (a ratite group), Megapodidae (mound builders), five families of parrots, Aegothelidae (owlet nightjars), Orthonychidae (logrunners), Pomatostomatidae (false babblers), Acanthizidae (Australian warblers), Maluridae (fairy wrens), Eopsaltriidae (Australian robins), Climacteridae (treecreepers), Meliphagidae (honeyeaters), Cracticidae (butcherbirds), Ptilonorhynchidae (bowerbirds), and Paradisaeidae (birds of paradise). Of these, the radiation of Paradisaeidae is spectacular in New Guinea, particularly in highlands.

The lowland rainforests of New Guinea support relatively large numbers of species of Accipitridae (hawks), Columbidae (pigeons), Psittacidae and Loriidae (parrots), Cuculidae (cuckoos), Alcedinidae (kingfishers), Campephagidae (cuckoo shrikes and trillers), Acanthizidae (Australian warblers), Rhipiduridae (fantails), Myiagridae (monarch flycatchers), Pachycephalidae (whistlers, shrike thrushes), Sturnidae

Table 5.1. Taxonomic Compositions of Rainforest Avifaunas in Lowland New Guinea at the Gogol–Naru Area (GR) and Brown River (BR, data from Bell, 1982a), and in Tropical Australia (Cape York Peninsula) North of Cooktown (CYP), Lowlands (LT) and Tablelands (TT) Between Townsville and Cooktown

Families	GR	BR	CYP	LT	TT
Casuariidae (cassowaries)	1	1 (1)	1 (1)	1 (1)	1 (1)
Ardeidae (herons)	3	1	2 (2)	0	0
Accipitridae (hawks, eagles)	8	13 (7)	9 (8)	3 (2)	2 (2)
Falconidae (falcons)	1	1	1 (1)	0	0
Megapodidae (mound builders)	2	2 (1)	2 (1)	2 (1)	1
Rallidae (rails)	2	3 (2)	2 (2)	2 (2)	0
Columbidae (pigeons, doves)	19	27 (18)	8 (5)	8 (5)	7 (4)
Cacatuidae (cockatoos)	2	2 (2)	3 (2)	1 (1)	1 (1)
Psittacidae (parrots, pygmy parrots)	4	6 (4)	2 (2)	0	0
Loriidae (brush-tongues)	6	7 (4)	2 (1)	2 (1)	2 (1)
Opopsittidae (fig parrots)	2	1	1 (1)	1 (1)	1 (1)
Polytelitidae (king parrots)	0	1	2	1	1
Platycercidae (rosellas)	0	0	0	1	1
Cuculidae (cuckoos)	7	11 (7)	9 (5)	7 (4)	4 (2)
Strigidae (owls)	1	3 (1)	3 (1)	3 (1)	3 (1)
Tytonidae (barn owls)	0	1	1	1	1
Podargidae (frogmouths)	1	1 (1)	3 (1)	1 (1)	0
Aegothelidae (owlet nightjars)	0	1	0	0	1
Caprimulgidae (nightjars)	1	2 (1)	1 (1)	1 (1)	1 (1)
Hemiprocnidae (tee-swifts)	1	1 (1)	0	0	0
Apodidae (swifts)	2	4 (2)	1	1	1
Alcedinidae (kingfishers)	10	11 (9)	8 (6)	4 (3)	2 (1)
Meropidae (bee eaters)	1	1 (1)	1 (1)	1 (1)	1 (1)

Table 5.1. (*Continued*)

Families	GR	BR	CYP	LT	TT
Coraciidae (rollers)	1	1 (1)	1 (1)	1 (1)	1 (1)
Bucerotidae (hornbills)	1	1 (1)	0	0	0
Pittidae (pittas)	2	2 (2)	2 (1)	1	1
Hirundinidae (swallows)	1	1 (1)	1	1	1
Campephagidae (cuckoo shrikes, trillers)	5	6 (3)	5 (3)	4 (3)	4 (3)
Turdidae (thrushes)	0	0	0	1	1
Orthonychidae (logrunners, whipbirds)	1	1 (1)	0	1	2
Pomatostomatidae (false babblers)	1	1 (1)	0	0	0
Maluridae (fairy wrens)	0	3	1	1	1
Acanthizidae (Australian warblers)	5	3 (3)	3 (2)	5 (2)	8 (2)
Rhipiduridae (fantails)	5	6 (4)	3 (1)	2	2
Myiagridae (monarch flycatchers)	8	6 (5)	9 (3)	5 (2)	2 (1)
Eopsaltriidae (Australian robins)	3	3 (2)	5 (2)	1	1
Pachycephalidae (whistlers, shrike thrushes)	6	4 (3)	3 (2)	2 (2)	3 (1)
Climacteridae (treecreepers)	0	0	0	0	1
Dicaeidae (flowerpeckers)	2	2 (2)	1	1	1
Nectariniidae (sunbirds)	2	2 (2)	1 (1)	1 (1)	0
Zosteropidae (white-eyes)	0	1	1	1	1
Meliphagidae (honeyeaters)	13	19 (12)	13 (4)	5 (2)	3
Estrildidae (grass finches)	1	3 (1)	2 (1)	2 (1)	1
Sturnidae (starlings, mynas)	4	4 (4)	1 (1)	1 (1)	0
Oriolidae (orioles)	1	1 (1)	2	2	1
Dicruridae (drongos)	1	1 (1)	1 (1)	1 (1)	1 (1)
Cracticidae (butcherbirds, bell magpies)	3	3 (3)	1 (1)	1 (1)	1

Table 5.1. (*Continued*)

Families	GR	BR	CYP	LT	TT
Ptilonorhynchidae (bowerbirds)	1	1 (1)	3	1	4
Paradisaeidae (birds of paradise)	4	6 (3)	2 (2)	1	0
Corvidae	2	2 (2)	0	0	0
Total	147	184 (120)	123 (67)	83 (42)	72 (25)
% common with GR	–	65.0	–	–	–
% with GR and/or BR	–	–	54.5	51.0	35.0

Numbers in parentheses under the Brown River indicate the numbers of species shared with the Gogol–Naru area and those under Australian regions indicate the numbers of species shared with New Guinea sites.

(starlings and mynas), and Paradisaeidae (birds of paradise). The same applies to northern Australian rainforests except for the last two families, which are poorly represented in Australia.

The families not found in Australian rainforests are those of the Oriental region, which are represented by single species in New Guinea (Hemiprocnidae and Bucerotidae), and Pomatostomatidae (Australasian) and Corvidae (cosmopolitan), which have rainforest representatives in New Guinea but not in Australia. On the other hand, the families missing from the lowland rainforests of New Guinea are Platycercidae (Australian endemic), which has one rainforest species in northern Australia, and Turdidae (cosmopolitan) and Climacteridae (Australasian), which in New Guinea are restricted to highlands.

The rainforest of north Queensland tablelands contains several endemic species, all passerines, belonging to Acanthizidae (three species), Eopsaltriidae (one species), Pachycephalidae (one species), Orthonychidae (one species), Climacteridae (one species), Meliphagidae (one species), and Ptilonorhynchidae (two species). Most of them have vicariants in mid-mountain rainforest of New Guinea but not in subtropical rainforest of Australia (Kikkawa et al., 1981).

Habitat Characteristics and Niche Occupation

The rainforest habitat is most complex on the lowlands of New Guinea where high rainfall, coupled with soils rich in calcium and phosphorus, support multistoreyed tall vegetation that is characterized by irregular, closed canopy with scattered emergents, plank buttresses, robust woody lianes, vascular epiphytes, zingibers, aroids, and palms.

At an altitude of 30 m, Brown River lies on the Vanapa-Brown Rivers floodplain with moderately well-drained silty loam (Bell, 1982a). This relatively nutrient-rich soil and an annual rainfall of 1720 mm with a short dry season from July to September support a most developed structural type of rainforest, complex mesophyll vine forest (Webb et al., 1976). Bell's (1982a) site is located near the edge of unbroken rainforest (500 km²), which extends to highlands, and not very far from disturbed areas of rainforest and mosaic vegetation that border sclerophyll woodland of the drier region (Gillison,

1983). The site contains some deciduous trees, and leaf litter, which is normally absent or thin in layer, accumulates in the late dry season.

In the Gogol–Naru area (68,140 ha) topography varies from seasonally waterlogged floodplains (30 m in altitude) and river terraces to closely spaced, short ridges rising to heights of 150 m. The soils, derived from calcareous mudstone or alluvial deposits, consist of neutral medium clay rich in nutrients (Lamb, 1977). With a high annual rainfall of 3530 mm, complex mesophyll-macrophyll vine forest contains fewer deciduous elements than Brown River (Webb et al., 1976). The area has been logged extensively but is connected to highland rainforest without anthropogenic clearings.

The associated bird fauna at the Gogol–Naru area is one of the richest of northern lowland fauna within Papua New Guinea, and at Brown River it is the richest recorded per unit area anywhere in the tropics (Bell, 1982a). Their ecological compositions are expressed in niche occupation spectra (Kikkawa and Williams, 1971) in Figure 5.2.

Compared with the entire lowland land avifauna of New Guinea (Kikkawa and Williams, 1971), both Gogol–Naru and Brown River faunas have greater representations of arboreal herbivores (category 2, nectarivores/frugivores) and omnivores (category 3, nectarivores/frugivores/insectivores) and a smaller proportion of arboreal insectivores (category 1). Since these are typical trends of lowland bird communities in complex rainforest, the two communities would represent the most diverse fauna of the most complex rainforest in this region. The lowland trend is also seen in relatively large representations of small (category 4) and large (category 9) predators and the smallest representation of tree-nesting ground insectivores (category 5). All lowland trends, except small proportions of arboreal insectivores (category 1), are more pronounced at Gogol–Naru than Brown River. At Brown River facultative insectivores outnumber obligate insectivores generally, as noted by Bell (1982a). This fact may be related to the presence of a dry season when the upper layers of vegetation provide more fruits than insects. While this trend is strong in the arboreal insectivores (e.g., Acanthizidae, Myiagridae, Pachycephalidae), the ground insectivores (category 5), the smallest representation of all types, do not share the same trend with the arboreal insectivores. At Brown River, dense and clumped understorey cover provides shelter for ground-feeding, shrub-nesting insectivores (e.g., Maluridae, category 5) and herbivores (e.g., doves, category 6). Proportions of entirely ground-dwelling birds (category 7) and parasitic breeders (category 10) are similar between Brown River and Gogol–Naru. In Bell's (1982a) classification also, the percentage representations of obligate herbivores (frugivores, nectarivores, granivores), insectivores, mixed feeders (herbivores and insectivores), carnivores and omnivores (herbivores and carnivores) of the Gogol–Naru fauna are very similar to those of Brown River given by Bell (1982a).

In Australia the different faunal compositions of three subregions are also reflected in their niche occupation spectra (Figure 5.2). Each spectrum is constructed from a composite fauna of rainforest for the subregion.

The rainforest of Cape York Peninsula is largely of monsoon forest type developed under highly seasonal rainfall of up to 2049 mm (Iron Range) on lowlands and low hills. The largest patches of rainforest are located in the mid-peninsula region where McIlwraith Range rising to 823 m carries semideciduous mesophyll vine forest on deep soils or riverine alluvia of wet eastern slopes, notophyll vine forest with emergent *Araucaria* near the summit, and deciduous vine thicket on dry western slopes (Pedley

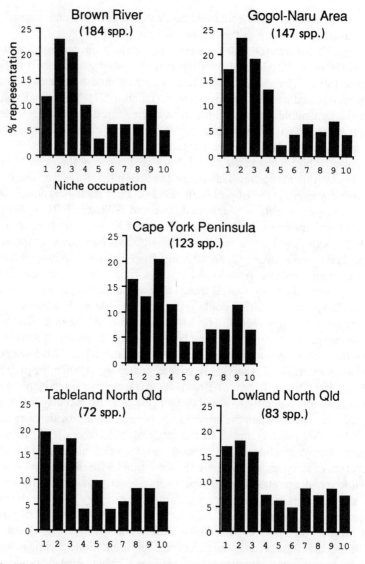

Figure 5.2. Niche occupation spectra of rainforest birds in New Guinea lowlands (Brown River and Gogol–Naru area), Cape York Peninsula, and tableland and lowland north Queensland, showing percentage representations of niche occupation types: 1, arboreal insectivores; 2, arboreal herbivores; 3, arboreal omnivores (mixed feeders); 4, small predators and scavengers. 5, ground insectivores nesting in vegetation; 6, ground herbivores nesting in vegetation; 7, ground-feeding, ground-nesting species; 8, aerial feeders; 9, large predators and scavengers; 10, parasitic breeders.

and Isbell, 1971). This rainforest area totals 96,590 ha (Winter et al., 1987). At Iron Range, strips of tall, semideciduous mesophyll vine forest with emergent *Ficus* occurs on lowland alluvia and semideciduous notophyll vine forest with sclerophyll emergents, grading into deciduous vine thicket on low hills below Mt. Tozer (545 m) (Pedley and Isbell, 1971). This patch is about 83,300 ha in area (Winter et al., 1987) and contains all rainforest bird species of New Guinea origin known from Cape York Peninsula. Other rainforest patches include 53,400 ha in the Jardine River catchment area and 9470 ha at Lockerbie in northern peninsula (Winter et al., 1987).

The niche occupation spectrum of rainforest birds of Cape York Peninsula (Figure 5.2) gives ecological characteristics of the attenuated New Guinea rainforest fauna diluted by the fauna of adjacent sclerophyll habitats. Thus, proportions of arboreal omnivores (category 3), large and small predators (categories 4 and 9), ground insectivores (category 5), aerial feeders (category 8), and parasitic breeders (category 10) are similar to those of Brown River, but both arboreal and ground herbivores (categories 2 and 6) are reduced in proportion. Particularly, arboreal nectarivores/frugivores are underrepresented in Cape York Peninsula rainforest. The strongly seasonal rainfall regime of Cape York Peninsula (Kikkawa et al., 1981) controls the distributions of impoverished rainforest floras in refugia (Webb and Tracey, 1981a), which do not produce plant food continuously throughout the year to support many herbivores. On the other hand, both the mosaic habitats and riverine rainforests support many sclerophyll habitat birds, particularly in the dry season. These are mostly insectivores belonging to Myiagridae, Eopsaltriidae and Pachycephalidae.

Within Australia the most optimum conditions of humid tropical lowland rainforest occur in tropical north Queensland south of Cooktown, where annual rainfalls exceed 2500 mm (Webb and Tracey, 1981a). Although temperatures fluctuate seasonally, nutrient-rich basaltic soils in high-rainfall areas (e.g., mean annual at Tully 4321 mm) support complex mesophyll vine forest similar in structure to lowland New Guinea rainforest. All life forms of New Guinea rainforest are well represented, but tree species diversity is not as great (Webb and Tracey, 1981a). Any extensive rainforest of this type in lowland tropical north Queensland is inhabited by all rainforest bird species known from this subregion. Their niche occupation spectrum most resembles that of Gogol–Naru though obligate and facultative nectarivores/frugivores (categories 2 and 3) and small predators (category 4) are very much reduced in proportion. Although plant food is utilized by more herbivores in proportion than in Cape York Peninsula rainforest, compared with Gogol–Naru the rainforest flora in humid lowland tropics of north Queensland is not as rich in plant food resources for birds (Webb and Tracey, 1981a, 1981b).

The niche occupation spectrum of tableland rainforest birds is included in the comparison (Figure 5.2) to show the effect of altitude in north Queensland rainforest birds that include several species endemic to the tablelands. The rainforest habitat is variable and life forms of lowland rainforest are not abundant. The most consistent difference is the smaller leaf size, characterizing notophyll vine forest. Structurally, it may be simple, mixed, or complex depending on the availability of soil nutrients. At higher altitudes the palms give way to tree ferns, and lichens on tree trunks to mosses. On mountain tops (e.g., Bellenden Ker, 1520 m) simple microphyll vine-fern thicket forms dense vegetation (Webb and Tracey, 1981a). As in other subregions, most rainforest

bird species are recurrent in different areas, but some endemic species are rare or localized in distribution. The lowland trends disappear from the niche occupation spectrum
of tableland birds though a substantial proportion of arboreal herbivores still remains.
The most conspicuous difference is in the proportion of insectivores, both arboreal
(category 1) and ground-feeding (category 5), which frequent dense thickets or the
understorey of tableland rainforest. The spectrum resembles that of low montane
(1000–1500 m) fauna of New Guinea (Kikkawa and Williams, 1971), except for ground
insectivores (category 5) and aerial feeders (category 8), which have greater representations in north Queensland tablelands.

Variation in the niche occupation spectrum may be considered to result from
responses of birds to the structural differences of vegetation along environmental
gradients (Kikkawa, 1988). Thus the lowland trends seen in the lowland New Guinea
rainforest and to some extent repeated in the rainforest of tropical humid lowlands of
north Queensland reflect resource availability in the most complex type of rainforest in
Australasia.

Species Overlap and Diversity Between Habitats

The degree of species overlap and between-habitat diversity (β diversity) were calculated between plots within each area to see how local habitat differences, particularly
those caused by disturbances, contributed to the diversity of the total fauna. For species
overlap the percentage of species common to two plots was obtained whereas for β
diversity, an index $(1 - C'\lambda)$ was calculated from the number of individuals of each
shared and unshared species between two samples (see Morisita, 1971). The index
takes 0 or small negative values for complete overlap and unity for no overlap (Kikkawa, 1986). All other diversity values were calculated according to the Brillouin's
equation for H with $\log_2$.

Gogol–Naru Area

A plot was situated in each of three major forest types (Whyte, 1975): forest on well-
drained terraces (terrace forest), hill forest, and floodplain (swamp) forest. Three
regrowth plots were in logged areas where each of the major forest types once existed.
Intensive logging began in 1973 and the regenerating vegetation on the logged plots was
no more than 5 years old except on a section of the hill regrowth plot where the forest
had been manually cleared for village gardens 10–15 years earlier. An additional forest
plot was located in the village reserve at Oupan. It contained a combination of hill and
terrace forest types.

Details of variation in habitat structure between the plots is given in Driscoll (1985,
1986). Briefly, the contrast between regrowth and forest vegetation can be defined by
the greater canopy height and development of tree layers at forest sites. Cauliflory,
epiphyllae, rough bark, plank buttresses, and palms were all typical features of
unlogged forest that were lacking or scarce in areas of regrowth. Regrowth trees were
typically smooth-barked with shallow crowns and large leaves, and vines were more
abundant. Foliage and litter tended to occur in clumps in the regrowth partly as a result

of ponding of water in old dozer tracks, a common feature of logged areas on flat terrain. Meijer (1970) suggests that heavy logging equipment used in Malaysian rainforests can cause soil disturbance and disappearance of seedlings over as much as 40% of the logged areas. The effect was of the same magnitude in the Gogol–Naru area, where the extent of timber extraction was often higher than in the example used by Meijer.

Considerable variation in habitat structure was not directly related to the effects of logging or the natural occurrence of treefalls (Driscoll, 1985, 1986). For example, clumping of shrub foliage, stilt roots, and a predominance of trees with flaky and scaly bark were additional features of the swamp forest plot, whereas smaller diameter trees with deeper crowns and less buttressing were typical of hill forest sites. Retention of soil moisture and slightly higher rainfall closer to the coast where the swamp forest plot was situated were important factors on low ground, whereas the effect of slope in shaping the crowns of canopy trees and causing higher rates of nutrient loss and treefall caused variation in habitat structure on hilly terrain.

Canopy Birds

An assessment of the diversity, relative densities, and vertical distribution of canopy species is made from data collected at two treetop platforms, one in the logged (regrowth) and the other in the unlogged forest plot on river terrace. Counts of bird species were made by scanning 180 degrees for 5 minutes and then turning to scan the opposite side for 5 minutes. The sequence was continually repeated, with birds being recorded for more than one 5-minute interval if they remained in view or came back into view. However, a bird known to have moved back and forth into view during the same 5-minute interval was counted only once. The purpose was to gain an index of relative density rather than estimate actual numbers of birds. There was adequate visibility from both platforms and birds were recorded as flying or perched in various height and distance zones (see Driscoll, 1985). Data were collected for 12 days at each plot (7 hours each day) between September 1978 and October 1979.

Results are shown in Table 5.2. For similar numbers of counts the forest plot had nine species more than the regrowth plot and 70% of the unshared species. Although there was a 75% species overlap between samples from the two plots, the counts of many species differed considerably as indicated by a moderately high β diversity of 0.40. Furthermore, counts of the more common species at the regrowth plot tended to be higher than counts of common species at the forest plot, giving species diversity values (H) of 4.57 and 5.15, respectively. At the regrowth plot, 50% of total counts were of the five most common species, whereas counts for a minimum of nine species from the forest plot need to be tallied to yield 50% of total counts.

Twenty-three canopy species were definitely more numerous at the forest plot compared with 13 that were more numerous at the regrowth plot. Counts of the remaining species were either similar for both plots or the data were too few to indicate a difference. The 23 species more numerous at the forest plot included four pigeons, two cockatoos, two parrots, two lorikeets (brush-tongues), two fig parrots, two cuckoos, a swiftlet, a cuckoo shrike, a robin, a whistler, a flowerpecker, two honeyeaters, a myna, and a butcherbird. The 13 species that were more numerous at the regrowth plot

Table 5.2. Results of Canopy Bird Counts in Unlogged (Forest) and Logged (Regrowth) Plots at Gogol–Naru

| | Plot | |
Result	Forest	Regrowth
No. of counts	8175	8199
No. of species	86	77
No. of shared species	70 (75%)	
Mean no. of species counted per day[a]	48	41

[a] Difference being significant at $p < 0.02$ (Mann-Whitney $U = 17.5$, d.f. = 22).

included one pigeon, two lorikeets (brush-tongues), a cuckoo, a tree-swift, a roller, a cuckoo shrike, a warbler, a monarch flycatcher, a starling, a butcherbird, and two birds of paradise.

The two groups of species outlined above can be compared on the basis of food preference (see Bell, 1982a). Of the "regrowth" species only one was an obligate herbivore, there being equal numbers of mixed feeders and obligate insectivores among the remaining 12 species. In contrast, nearly half (11) of the "forest" species were obligate herbivores, eight were mixed feeders, and four were obligate insectivores. Many of the obligate herbivores in the canopy at the forest plot were large birds in contrast to the generally smaller species typical of the regrowth canopy. A comparison of niche occupation types between the two groups of species also reveals important differences. The respective numbers of species of various categories for the forest and regrowth are 4:4 (category 1), 12:4 (category 2), 3:3 (category 3), 3:0 (category 4), and 1:2 (category 8). A much smaller representation of arboreal herbivores (category 2), absence of small predators (category 4), and increase in aerial feeders (category 8) utilizing open canopy all point to disturbances and weakening of lowland trends discussed in Section 3 at the regrowth plot.

Understorey Birds

Data for understorey species were collected through the systematic use of mist nets. Nets (12×1.8 m or 9×2.3 m with mesh size of 32 mm) were set at ground level and all captured birds were banded, measured, and released. On the seven study plots a total of six rounds of netting were conducted over a period of 10 months from June 1977. Nets were set at each plot for 2 days at a time (approximately 8 hr 45 min per day). For the first two rounds of netting, two lines of five nets were used and an additional line of five nets was added for subsequent rounds. All adjacent sites were roughly 60 m apart yielding plot dimensions of 270 m × 110 m or an area of slightly less than 3 ha. Netting was continued on two plots on river terrace (forest and regrowth). Seven extra nets sites were placed irregularly among the 15 already in use and six more netting rounds conducted, five between early June and the end of November 1978 and one at the end of September 1979 after a break of nearly a year.

Between-Habitat Diversity. Table 5.3 gives the number of species captured at each plot and species overlap and β diversity between plots. Similar species richness in terrace

Table 5.3. Species Richness, % Overlap (right of diagonal) and β Diversity (left of diagonal) Among Mist-Netted Birds at Gogol–Naru

Plots	No. of Species (Total: 54)	G1	G2	G3	G4	G5	G6	G7
G1 Terrace forest	35		**65.1**	**73.2**	43.9	48.8	44.4	39.6
G2 Oupan Reserve	36	**0.08**		**56.5**	46.3	44.4	46.7	36.0
G3 Hill forest	36	**0.09**	**0.01**		46.3	47.7	**50.0**	38.8
G4 Swamp forest	24	0.32	0.22	0.30		39.5	45.9	47.4
G5 Hill regrowth	29	0.37	0.28	0.20	0.25		**51.3**	45.2
G6 Swamp regrowth	30	0.58	0.55	0.51	0.42	0.22		**55.0**
G7 Terrace regrowth	32	0.58	0.56	0.59	0.35	0.21	**0.02**	

Note: Pairs of plots with 50% or greater species overlap or with negligible β diversity are given in bold figures.

forest, Oupan Reserve, and hill forest, and very low β diversity between them testify to a basic similarity of the three communities. In contrast, the terrace and swamp regrowth bird communities are similar to one another and most different from those in forest plots with the exception of the swamp forest. Both the swamp forest and hill regrowth communities exhibit some similarities to other forest and regrowth plots, but for different reasons. Compared with other regrowth plots, logging on the hill regrowth plot had been very patchy and the resulting vegetation structure was very variable with enclaves of relatively undisturbed forest. These features explain a relatively high overlap of hill regrowth with the more typical bird communities of forest (Driscoll, 1986).

Of the species characteristically associated with relatively undisturbed, well-drained forest on level terrain, there are as many as 20 understorey species that were at the highest densities in the terrace forest plot, but were slightly less abundant on hillsides and least abundant or absent from swamp forest and regrowth vegetation (Driscoll, 1985, 1986). The capture rate of birds was also significantly different between plots, being as much as 40% lower in both regrowth vegetation and swamp forest compared with terrace forest (Driscoll, 1985).

The swamp forest community lacked certain species found at other forest sites and also lacked a large contingent of transitory species, which was a feature of regrowth vegetation (see below). The lowest number of species was recorded at the swamp forest plot; a third lower than at other forest plots. Bird communities in regrowth vegetation and swamp forest were similar because of common absences of more typical forest birds (Driscoll, 1985) but also because of an overlap of other species. A monarch flycatcher (*Myiagra alecto*) and a fantail (*Rhipidura leucothorax*) were especially abundant in regrowth vegetation and the latter species was also common in swamp forest (Driscoll, 1985). The similarity between swamp forest and regrowth communities was further strengthened by low numbers or the absence of many species typical of terrace forest, and the occurrence of three species of kingfisher (*Ceyx azureus*, *C. pusillus*, and *Halycon nigrocyanea*). A paradise kingfisher (*Tanysiptera galatea*), typical of terrace and hill forest, was rarely recorded from swamp forest and regrowth (Driscoll, 1985).

There was a core of species that did not appear to be affected by logging, once the regrowth became established, and ranged widely even though they were not uniformly conspicuous. Their distributions and local patterns of activity were presumably related to features of the environment relatively unaffected by logging or natural disturbances (Driscoll, 1986).

Species Richness and Irregular Occurrence of Species. In considering data from the two plots on river terrace where most sampling was conducted, the two species *Rhipidura leucothorax* and *Myiagra alecto* dominated the catch from the regrowth plot. These species accounted for 28% of the individuals banded (n = 193), and were recorded from at least two-thirds of the sites and at least 10 out of the 12 episodes of netting. The species netted most often at the forest plot (*Monarcha guttula* and *Toxorhamphus novaeguineae*), on the other hand, accounted for 20% of the individuals banded (n = 290). Altogether, about the same number of species were banded at the two plots on river terrace; 41 at the forest plot and 42 at the regrowth plot. On average, however, two-thirds the number of species were taken at a single net site or during a single netting episode at the regrowth plot, the differences being statistically significant (Driscoll, 1985). Thus, the total number of birds netted at the forest plot was much greater (290) than at the regrowth plot (193), resulting in greater species diversity (H) for the forest plot (4.45) than the regrowth plot (4.21). In considering individual species, many were being caught irregularly at the regrowth plot, irregular in the sense of only being caught at one or two net sites or in one or two netting episodes. The number of species in either of these categories at the regrowth plot was twice (22) that at the forest plot (10). These differences between the plots were probably due to both greater spatial and temporal variation in vegetation structure at the regrowth plot and to the more sporadic occurrence of species in the regrowth. For instance, records of *Rhipidura rufidorsa*, *Tanysiptera galatea*, and *Crateroscelis murina* were of single juveniles. Similarly, *Pomatostomus isidorei*, *Ptilorrhoa caerulescens*, *Chalcophaps stephani*, and *Monarcha guttula* did not appear to be truly established in the regrowth, although the few birds netted there were adults.

In general, individual birds were remaining at the forest plot for longer periods (resident) than at the regrowth plot, where only one species (*Rhipidura leucothorax*) was represented by individuals caught on at least four occasions. Ten species were in this category at the forest plot. Nevertheless, at both sites a high proportion of captured birds was made up of transients (Driscoll, 1985).

Vertical Stratification of Species

The majority of species remained within limited height intervals. Only 24 species were both counted and netted at the two terrace plots, representing 26% of species counted and 44% of species netted. Of the 54 species netted at either plot, 41% were obligate insectivores, 17% were carnivores, 11% were obligate herbivores. In contrast, of those species (93) counted from either platform, obligate herbivores (32%) outnumbered obligate insectivores (21%), and carnivores (9) were not as well represented as in the understorey. The proportion of species that were mixed feeders was slightly higher for the canopy than for the understorey, 35 and 28%, respectively. Omnivores constituted only about 3% of the species from either data set.

At the forest plot, only 16 species were captured in ground nets and counted from the platform, and only four of these were both netted and counted in appreciable numbers.

The four species, *Meliphaga* sp., *Xanthotis flaviventer*, *Pitohui kirhocephalus*, and *Dicrurus hottentottus*, were all mixed feeders. At the regrowth plot, 19 species were both netted and counted, 10 of which came under the same category for the forest plot. The data do not indicate that any of the latter 10 species, upon inhabiting the regrowth, had altered the general height at which it foraged.

Some insectivores and mixed feeders were regularly feeding in the outer layer of foliage, which included the regrowth canopy and contiguous foliage supported by remnant tall trees. As a consequence, a bird of paradise (*Paradisaea minor*) and a honeyeater (*Glycichaera fallax*) were coming much closer to the ground at the regrowth plot where they were taken in ground nets. The honeyeater was netted four times at the regrowth site but only once at the forest plot. In contrast, it was counted more often from the forest platform than from the regrowth platform.

At the regrowth plot, the flowerpecker *Dicaeum pectorale* was captured once, and the mannikin *Lonchura tristissima* was counted once. These records suggest that both species were adhering less rigidly to their characteristic foraging zones than they would in their normal habitats. This flowerpecker was a canopy species whereas the mannikin foraged mostly in roadside grasses.

Additional information is available on the vertical distribution of species that were counted 10 times or more in perched positions. Counts of these birds were divided into those above and below 35 m for the forest plot and those above and below 25 m for the regrowth plot. These heights represent the most even division of total counts at the respective plots. Species generally fall into one of three categories: they either clearly occupy one or the other height zone, or exhibit no preference.

Forest Plot. Among 11 obligate frugivores at the forest plot, four columbids— *Ptilinopus pulchellus, P. nanus, P. magnificus,* and *Ducula zoeae*—were counted in the canopy at heights below 35 m far more often than in emergent vegetation. In contrast, the largest arboreal feeding columbid, *Ducula pinon*, perched more often in vegetation above 35 m. Similarly, *P. perlatus*, the largest species of fruitdove, and *P. iozonus* were seen more often high in the canopy. Like *D. pinon*, the latter two species generally foraged in groups of at least six birds. The frugivorous cuckoo shrike *Coracina boyeri* also moved about in groups and tended to occupy the tallest vegetation, as did the channel-billed cuckoo *Scythrops novaehollandiae*. Counts of the two mynas (*Mino anais* and *M. dumonti*) were fairly evenly distributed between the different levels of the canopy.

Two large herbivores not restricted to frugivory, the cockatoos *Cacatua galerita* and *Probosciger aterrimus*, differed in the zone they most frequented. *P. aterrimus* was invariably seen in canopy below 35 m whereas *C. galerita* perched at heights above 35 m. Other herbivores either confined their activities to below 35 m or were active in both height zones.

Three obligate insectivores, *Eurystomus orientalis, Peltops blainvillii* and *Microeca flavigaster*, primarily hawked for insects from exposed perches and were mostly counted in the zone above 35 m. These three species varied considerably in size, weighing approximately 126 g, 30 g (Bell, 1982b), and 15 g (estimated), respectively. Another obligate insectivore, the bronze cuckoo *Chrysococcyx meyerii*, was also counted more often above 35 m. The flycatcher *Monarcha chrysomela* and whistler

Pachycephala monacha were counted about the same number of times above and below 35 m. The small insectivorous honeyeater *Glycichaera fallax* tended to keep lower in the canopy and, as already noted, sometimes visited the shrub layer, at least at the regrowth plot.

The omnivore *Cracticus cassicus* was normally seen in the highest zone at both plots. Similarly, the cuckoo shrike *Coracina papuensis* and two small species, the sunbird *Nectarinia aspasia* and the honeyeater *Oedistoma pygmaeum*, were the only mixed feeders counted predominantly above 35 m. Other mixed feeders were rarely counted above 35 m and normally foraged in the middle storey.

Regrowth Plot. With the exceptions noted below, species at the regrowth plot were counted most often between the heights of 25 and 35 m. Eighty-four percent of perched counts ($N = 5741$) were made in the 25- to 35-m zone where the crowns of trees surviving the logging were at a density of 18 per ha (57 crowns within a 100-m radius). In the 15- to 25-m zone, the surviving trees were at a slightly higher density of 22 per ha, but the total mass of foliage was much less than in the zone above because the crowns of smaller trees were far less extensive. Only in a few scattered locations did foliage extend above 35 m. The predominance of counts in the zone between 25 and 35 m was primarily due to the presence of a greater volume of vegetation serving as foraging substrate. Also, the height of the zone may have offered easy access to those species accustomed to moving among the canopy and emergent trees of unlogged forest.

Local foci of bird activity were not only apparent as a response to the vertical concentration of foliage in the regrowth but also as a response to horizontal placement of remnant tall trees. Within 100 m of the platform, 61% of tall trees were located in the northern half of the field of view, which was the same as the percentage of bird counts made on that side.

At the regrowth plot, *Ptilinopus magnificus* was the only obligate frugivore counted more often below 25 m (89% of counts). Similarly, at the forest plot it was most active in the subcanopy. Other frugivores counted in reasonable numbers at both plots tended to occupy a narrower vertical range at the regrowth plot, being seldom seen below 25 m or above 35 m. Apart from *Ptilinopus magnificus* two other obligate herbivores, the lories *Pseudeos fuscata* and *Lorius lory*, were recorded most often below 25 m at the regrowth plot. These lories concentrated their activity on a small, flowering umbrella tree.

Of nine insectivores counted from the platform at the regrowth plot, only three species were counted on more than 20% of occasions below 25 m. Seventy-eight percent of counts of the flycatcher *Monarcha chrysomela* were of birds below 25 m where they foraged across the top of the regrowth canopy, often as members of mixed foraging flocks. Similarly, about 50% of counts of a warbler (*Gerygone magnirostris*) and an insectivorous honeyeater (*Glycichaera fallax*) were recorded for the zone below 25 m because these species utilized the regrowth canopy and vegetation contiguous with it on remnant trees. Even though mixed feeders at the forest plot mostly utilized only the lower height zone, at the regrowth plot only one mixed feeder, the honeyeater *Xanthotis flaviventer*, was counted most often below 25 m. Nevertheless 20–40% of counts of six other mixed feeders were of birds below 25 m.

Table 5.4. Total Species Richness, % Overlap (right of diagonal), and β Diversity of Mist-Netted Understorey Species (left of diagonal) Among 4-ha Plots of Lowland Rainforest in Cape York Peninsula North of Cooktown

Plots	Latitude (S)	No. of Species (Total: 75)	No. of Species Mist-Netted (Total: 31)	Plots				
				C1	C2	C3	C4	C5
C1 Lockerbie	10°47'	39	12		**50**	**65**	**60**	45
C2 Lake Boronto	10°46'	39	13	0.21		**52**	47	42
C3 Capelands	11°40'	, 40	11	0.09	0.09		**56**	49
C4 Iron Range	12°45'	46	17	0.26	0.35	0.26		44
C5 McIvor R./ Mt. Webb	15°05'	42	10	0.13	0.22	0.01	0.38	

Note: Pairs of plots with 50% or greater species overlap are given in bold figures.

Cape York Peninsula

For the comparison of communities within Cape York Peninsula data from five rainforest plots were used (Table 5.4). Lockerbie (C1) is in the northernmost patch of well-developed rainforest on the continent, which supports semideciduous mesophyll vine forest to an elevation of 150 m on deep red and brown soils derived from Mesozoic sediments under a seasonal rainfall regime. Lake Boronto (C2) and Heathlands (C3) are small patches of evergreen (simple) notophyll vine forest with sclerophyll (*Acacia*) emergents growing on deep sand (C2) or sandy loam on laterite (C3). C2 is located adjacent to coastal heath and lake shore whereas C3 is in the Jardine catchment area (150 m in altitude) and surrounded by heath and eucalypt forest (Laverack and Stanton, 1977). Iron Range (C4) combines tall, semideciduous mesophyll vine forest on the nutrient-rich alluvium of the Claudie River with semideciduous mesophyll vine forest containing sclerophyll (*Acacia*) emergents growing on adjacent low hills. The area is in the best developed and most extensive lowland rainforest patch in Cape York Peninsula. McIvor River and Mt. Webb (C5) are located to the south of the major geographic barrier of Cape York Peninsula (Kikkawa et al., 1981), where transitional rainforest fauna is very much impoverished and found in narrow strips of gallery forest (complex mesophyll vine forest) along rivers and small patches of semideciduous mesophyll vine forest on basaltic hills. C5 combines gallery forest of McIvor River and hill forest of Mt. Webb in this region.

The field data were collected between October 1974 and July 1977 by visiting each plot for 5 days twice, in wet and dry seasons, except for C5 which was visited once in the dry season only. All land birds recorded in 4-ha plots were used for the expression of species richness and overlap. For β diversity only the mist net data were used. The data consist of captures in 10 nets (each 13 × 2 m) set at the ground level over 5 mornings (6 hours from dawn) for two periods. In C5 the captures were made in the same total net-hours but in one period only.

Results in Table 5.4 show a great similarity of fauna in different patches of rainforest within Cape York Peninsula. The small differences are due to two factors. One is ecological in that C2 and C3 are structurally and floristically impoverished rainforest, in which the bird species adapted to complex mesophyll vine forest are poorly represented while species commonly found in sclerophyll habitats are recorded. Typical rainforest birds of Cape York Peninsula absent from these plots for ecological reasons are *Tanysiptera sylvia* (C3), *Pitta erythrogaster* (C2), *Drymodes superciliaris* (C2), *Tregellasia leucops* (C2), *Sericornis beccarii* (C2), and *Oriolus flavocinctus* (C3). The sclerophyll habitat or forest edge species recorded from these plots are *Geopelia humeralis* (C2), *Centropus phasianinus* (C2), *Caprimulgus macrurus* (C3), *Halcyon sancta* (C2), *Microeca griseoceps* (C2), *Myiagra rubecula* (C2), *Lichmera indistincta* (C2), and *Ramsayornis modestus* (C2). The other is biogeographic in that C4 has the richest rainforest fauna of the peninsula and this fauna is attenuated across the barrier both to the north (C1, C2, C3) and to the south (C5). Absent from these plots because of the attenuation are *Eclectus roratus* (C1, C2, C3, C5), *Geoffroyus geoffroyi* (C1, C2, C3, C5), *Psittaculirostris diophthalma* (C1, C2, C3), *Pitta erythrogaster* (C5), *Drymodes superciliaris* (C5), *Poecilodryas superciliosa* (C5), *Arses telescophthalmus* (C5), *Xanthotis flaviventer* (C5), *Ptiloris magnificus* (C5), and *Manucodia keraudrenii* (C5). Geographic factors are also responsible for the presence of *Xanthotis macleayana* only in C5.

Unlike at Gogol–Naru, habitat impoverishment in northern peninsula excluded mostly ground predators rather than frugivores whereas the geographic barrier excluded distribution of obligate and facultative herbivores as well as ground insectivores to the south.

Tropical North Queensland

In tropical north Queensland, vegetation structure changes along altitudinal, rainfall, and edaphic gradients from the most complex rainforest of humid lowlands (Webb, 1968; Tracey and Webb, 1975; Tracey, 1982). The bird species diversity and overlap were calculated from data collected in relatively undisturbed rainforests of representative types and in paired plots of contrast between disturbed and undisturbed rainforest of the same type.

Comparison Between Undisturbed Areas

For the comparison of undisturbed rainforests 10 plots were selected from both lowlands and tablelands (Table 5.5). They represent complex mesophyll vine forest on basaltic soil (Bailey's Creek on lowlands, McNamee Creek, Maalan, and Palmerston on foothills), mesophyll vine forest on lowlands (Lacey's Creek with some *Acacia* emergents, Cardwell with bordering eucalypt forests), mesophyll vine forest on tablelands alluvium derived from granite (Davies Creek), simple mesophyll-notophyll vine forest developing as an understorey of sclerophyll woodland on lowlands (Daintree), complex notophyll vine forest of tablelands (Crator), and simple notophyll vine forest of tablelands (Gurkha Pocket) (Webb, 1966; Tracey, 1982). Each plot was approximately 4-ha in size and rectangular in shape, with one side following a track or stream for a distance of 270 to 400 m. Birds were censused twice at the height of the breeding season (October to December) between 1966 and 1968. Complete species lists for three of the plots are given in Kikkawa (1968). All species were identified and individuals were counted from the songs and calls heard during 1.5–2.0 hours from dawn. Some species

Table 5.5. Species Richness and Diversity of Birds in 4-ha Plots of Tropical Rainforests in North Queensland, Based on Dawn Census in the Breeding Season

Plots	Habitat[a]	Total no. of Species (Total: 85)	No. of Strictly Rainforest Species (Total: 57)	No. of Individuals Heard	Diversity (*H*)
Lacey's Ck	LL/MVF	31	30	97	3.98
Bailey's Ck	LL/CMVF	33	29	93	3.99
Palmerston	LL/CMVF	43	37	81	4.27
McNamee Ck	LL/CMVF	35	31	64	4.01
Cardwell	LL/MVF•SL	36	24	85	4.08
Daintree	LL/SL•MNVF	36	22	76	4.02
Crator	TL/CNVF	34	28	70	3.96
Gurkha Pkt	TL/SNVF	39	32	97	4.32
Davies Ck	TL/MVF	40	32	79	4.24
Maalan	LL•TL/CMVF	32	28	80	3.98

[a] LL = lowlands, TL = tablelands, MVF = mesophyll vine forest, CMVF = complex mesophyll vine forest, SL = sclerophyll elements, MNVF = mesophyll-notophyll vine forest, CNVF = complex notophyll vine forest, SNVF = simple notophyll vine forest.

called only once or twice while most others called frequently for about 20 minutes during the first hour of recording. A Uher tape recorder with a parabolic reflector was used to record vocalizations in the census. The method produced repeatable results, but abundance was considered relative and could only be used for comparison between similar habitats. For most species less than 50% of individuals present in the area would have been recorded. However, all except nocturnal species in the area were probably recorded by this method. In another study, mist netting in 18 areas of north Queensland revealed no species additional to those recorded during the dawn census (Kikkawa, 1982).

Results presented in Table 5.5 show that, despite the large total number of species recorded, species richness, density, and diversity varied little from plot to plot. The number of species recorded ranged from 31 to 43 among the lowland plots and 34 to 40 among the tableland plots. The number of strictly rainforest species varied more for the lowland plots (22–37) than for the tableland plots (28–32). The number of individuals ranged similarly among the lowland plots (64–97) and the tableland plots (70–97). Consequently, species diversity (*H*) had similar values and overlapped greatly between lowland plots (3.98–4.27) and tableland plots (3.96–4.32). These results seem to indicate that, unlike New Guinea lowlands, Australian rainforests do not support a high density or diversity of birds per unit area. Also, not all locally available species were recorded within the area of 4 ha. For example, a later study at Lacey's Creek (Kikkawa, 1982) recorded 63 species for the general area but still only 32 species in a 4-ha plot used for mist netting. Crome's (1978) record of 74 species for his study area at Lacey's Creek included many nonrainforest birds, and altogether the total number of species known from this area now stands at 92.

The largest number of nonrainforest species was recorded at Daintree (12 species) where *Acacia* and *Melaleuca* formed the canopy above the developing rainforest and attracted many sclerophyll habitat species. In other plots many birds had a wide ecological distribution, particularly in the coastal complex habitats, while very few forest

Table 5.6. Percentage Overlap of Species (right of diagonal) and β Diversity (left of diagonal) Among 4-ha Plots of Tropical Rainforests in North Queensland, Based on Dawn Census in the Breeding Season

Plots	N1	N2	N3	N4	N5	N6	N7	N8	N9	N10
N1. Lacey's Ck		**63**	**57**	**56**	35	33	24	28	28	30
N2. Bailey's Ck	**0.03**		**58**	**55**	47	44	26	33	30	27
N3. Palmerston	0.16	**0.06**		**63**	39	34	33	44	41	29
N4. McNamee Ck	0.18	0.14	**0**[a]		37	42	33	45	39	34
N5. Cardwell	0.15	**0.07**	0.16	**0.05**		39	23	23	25	21
N6. Daintree	0.39	0.16	0.37	0.32	0.14		21	23	19	13
N7. Crator	0.60	0.55	0.37	0.39	0.58	0.74		**59**	**61**	**50**
N8. Gurkha Pt	0.57	0.58	0.32	0.31	0.69	0.76	**0**		**68**	**61**
N9. Davies Ck	0.62	0.72	0.34	0.27	0.71	0.84	**0**	**0**		**64**
N10. Maalan	0.56	0.79	0.49	0.41	0.82	0.91	0.16	**0**	**0**	

[a] Small negative values due to high degrees of overlap are replaced with zero.
NOTE: Pairs of plots with 50% or greater species overlap or with negligible β diversity are given in bold figures.

edge species (e.g., *Malurus lamberti*, *Centropus phasianinus*, *Nectarinia jugularis*) were involved. Those, however, are largely the effect of natural habitat mosaic and transitions. Species replacement as a result of human disturbances will be discussed later.

The species overlap and β diversity between the plots are given in Table 5.6. As expected from the known dichotomy of distributional patterns between lowland and tableland faunas (Kikkawa, 1982), the species overlap between the rainforest plots of lowlands (C1–C4) and those of tablelands (C7–C10) was low, down to 24%, whereas within the lowlands (55–63%) and within the tablelands (50–68%) the overlap was no less than expected from minor differences between the rainforest habitats sampled. Similar trends were found in β diversity, which between the tableland plots became negligibly small.

The decrease of overlap and the increase of β diversity due to the inclusion of sclerophyll elements (C5, C6) in the rainforest were almost as great as shifting from lowland to tableland rainforests. Here the degree of overlap depended on the number of nonrainforest species drawn from a large pool of such species in the region. Therefore, the overlap between the areas of mixed vegetation is expected to be small.

Comparison Between Undisturbed and Disturbed Areas

Clearing of north Queensland rainforests occurred in both lowlands and tablelands during the past 100 years (Winter et al., 1987). Early clearing removed rainforests from fertile lands for sugarcane fields on the lowlands and cattle raising and farming on the tablelands. Today the remnants of rainforest occur in various sizes and shapes, and the boundaries of large intact areas present most irregular patterns (Tracey and Webb, 1975).

For comparison of bird fauna, four plots (two on lowlands and two on tablelands) were selected in such boundary areas where recent human disturbances have not eliminated the rainforest completely (Table 5.7).

Table 5.7. Percentage Overlap of Species and β Diversity of Understorey Species Based on Mist Net Data Between Undisturbed Rainforest and Disturbed Rainforest of Same Type in Tropical North Queensland

	Undisturbed Plot			Disturbed Plot			
Location	Total No. of Species	No. of Mist-Netted Species (Individuals)	Location	Total No. of Species	No. of Mist-Netted Species (Individuals)	% Overlap	β Diversity
Lacey's Creek (mesophyll VF)	63	12 (49)	Freshwater Valley (regrowth bordering sugarcane field)	62	25 (155)	60	0.57
Oliver Creek (complex mesophyll VF)	42	6 (23)	Noah Creek (cleared edge and cattle grazing)	41	7 (12)	42	0.32
Severin forest (intact) (complex notophyll VF)	60	20 (121)	Severin forest (recent selective logging)	56	27 (223)	79	0.00
"	"		Lake Eacham (tourist impact)	57	17 (67)	65	0.05

The Freshwater Valley plot was located at the edge of an extensive cane field and included highly disturbed rainforest on the creek bank and regrowth on the hillside. Noah Creek was a small area of recently cleared land within an extensive lowland rainforest. The plot was in bordering regrowth on the hillside and a narrow strip of rainforest along a creek, both highly disturbed by grazing and camping cattle.

On the tablelands, Severin forest was selected from a forestry area in which selective logging (up to 50% of canopy cover) had occurred recently. The plot was studied twice, 6 months and 12 months after the logging. Lake Eacham was selected for the impact of tourism, which in recent years created disturbances to the rainforest bordering picnic and parking areas. Lake Eacham is one of the most frequently visited tourist destinations on tablelands.

All plots were visited twice, in summer (wet season) and winter (dry season), and were mist-netted at 10 sites for five mornings on each occasion, except for Noah Creek where netting was not possible during the winter visit. This study formed part of a larger project carried out between 1973 and 1976, and results including ecological relationships of bird communities between these and other areas of north Queensland were discussed elsewhere (Kikkawa, 1982). Species overlap and β diversity between these plots and control plots were calculated for the present analysis. These are presented in Table 5.7, together with the basic data for each plot.

Human impact on rainforest was greater in the lowland plots than the tableland plots. There seem to be two reasons for this. First, there are more nonrainforest species on the lowlands available to colonize and utilize edge conditions created by the disturbance. Second, the features of complex rainforest that in intact areas produced the "lowland trends" of the niche occupation pattern disappeared as a result of disturbances.

At Freshwater Valley typical woodland species, such as *Geopelia placida*, *Pachycephala rufiventris*, *Grallina cyanoleuca*, and *Emblema temporalis*, and an edge species, *Malurus lamberti*, were recorded from the edge. Missing species included frugivores (*Ptilinopus regina*, *Lopholaimus antarcticus*, *Coracina lineata*) and large ground birds (*Casuarius casuarius*, *Megapodius reinwardt*). Numerical responses to edge conditions contributed to a high β diversity with the control plot. They included flocking *Zosterops lateralis* visiting shrubbery and concentration of lowland insectivores and mixed feeders (*Pachycephala simplex*, *Gerygone palpebrosa*, *Meliphaga notata*, and *M. gracilis*) in the understorey. At Noah Creek, although open forest species (*Halcyon macleayii*, *Artamus leucorhynchus*) and edge species (*Myiagra alecto*) were present, very few frugivores or ground predators were missing. The smallness and recency of clearing seem to be preventing differentiation of the fauna of this area to a greater extent than is the case for Freshwater Valley.

On tablelands selective logging appears to have produced little change in the fauna. Absences of few typical tableland rainforests species from either the impact or the control plot seem to be due to errors of sampling. The only possible change due to logging is the absence of *Pitta versicolor* (ground bird) in the logged plot. However, there is a pronounced numerical response of understorey insectivores (*Poecilodryas albispecularis*, *Sericornis magnirostris*, *S. citreogularis*, *Rhipidura rufifrons*, *R. fuliginosa*) to the openings in the forest. Increase in the density of these species in the logged plot, however, did not contribute to β diversity to any appreciable degree. Disturbances caused by tourist impact on tablelands was of a different nature. It attracted *Platycercus elegans*, a granivore, to the area and reduced the number of understorey birds. The absence of *Orthonyx spaldingii* from Lake Eacham may be due to a historical factor, but the low densities of *Sericornis* spp. and *Poecilodryas albispecularis* may reflect low availability of insect food in the understorey. These differences were not significant enough, however, to influence β diversity with the control plot.

Summary and Discussion

Species Diversity and Composition

The total number of bird species found in lowland rainforest is much greater in Central and South America and Africa than in New Guinea (Bourlière, 1983). The Australian rainforest contains even fewer species. However, the Australasian rainforest species belong to at least 40 families (50 families in the classification used in Table 5.1), which is comparable to Malaysian, African, or South American rainforest. This suggests less species packing in major families and possibly more specializations in minor families in the Australasian rainforest avifauna. Only a few families are represented by more than five species. They include Accipitridae (hawks), Columbidae (pigeons), parrot families, Cuculidae (cuckoos), Alcedinidae (kingfishers), Campephagidae (cuckoo shrikes), Acanthizidae (Australian warblers), Rhipiduridae (fantails), Myiagridae (monarch flycatchers), Pachycephalidae (whistlers), Meliphagidae (honeyeaters), and Paradisaeidae (birds of paradise). In a Malaysian rainforest, well-represented families are Cuculidae, Capitonidae (barbets), Picidae (woodpeckers), Campephagidae (minivets), Pycnonotidae (bulbuls), Aegithinidae (leafbirds), Timaliidae (babblers), Muscicapidae (Old World flycatchers), Sylviidae (warblers), Dicaeidae (flower-peckers), and Nectariniidae (sunbirds) (McClure, 1967). In the rainforest in the lower Amazon many species of Trochilidae (hummingbirds), Picidae (woodpeckers), Furnariidae (spinetails), Formicariidae (ant birds), Pipridae (manakins), Tyrannidae (tyrant flycatchers), and Thraupidae (tanagers) are abundant (Lovejoy, 1975). Such differences in taxonomic composition indicate the radiation of different groups in the rainforests of different regions.

The most complex community of New Guinea and tropical lowlands of Australia is characterized by arboreal nectarivores/frugivores, mixed feeders, insectivores, small predators, aerial feeders, and specialized ground feeders in the order of species abundance. This is called the lowland trend as the proportions of species occupying these niches change with altitude. Pigeons, parrots, kingfishers, and honeyeaters lose species while warblers gain species with altitude. In Costa Rica corresponding changes from the rich lowland fauna to the highland fauna are seen in the replacement of manakins, ant birds, cotingas, puffbirds, and motmots by warblers, vireos, and thrushes (Orians, 1969). Similar trends also appear along other environmental gradients in the tropics as associated structural differences of rainforest vegetation cause some divergence of niche occupation pattern shown by birds (Kikkawa, 1988). However, species overlap between rainforest plots of tropical lowlands remained greater than 50% in most comparisons. Only in swamp forest (Gogol–Naru) and mosaic vegetation with strong sclerophyll elements (north Queensland) did overlap with other types of rainforest fall below 50%. Otherwise the habitat differences caused no less species overlap than between plots of the same rainforest types within each region. Geographic effects (within Australia) appeared, however, in the degree of overlap, and these were more pronounced in numerical responses than species overlap as seen in β diversity values between Iron Range and other plots in Cape York Peninsula and between lowland and tableland plots in north Queensland.

There has been considerable speculation as to whether irregular occurrences of tropical species are related to large home ranges that are not adequately sampled in study plots of a few hectares, or whether individuals of a species are patchily distributed, possibly in relation to local heterogeneity of habitat (MacArthur et al., 1966; Terborgh and Weske, 1969; Karr and Roth, 1971; Willson et al., 1973; Lovejoy, 1975; Roth, 1976; Willson and Moriarty, 1976; Kikkawa et al., 1980). The netting results at Gogol–Naru revealed that individuals of some species moved in very confined areas but could either be packed close together or occur in isolation. However, many species obviously roamed over large distances, even in the understorey, and some individuals were recaptured after long periods of apparent absence. As pointed out by Karr (1975), the large number of bird species in rainforest is related to the great variety of ways in which food is available rather than by the high productivity of rainforest. Thus the average density of species is lower in tropical than in temperate communities (MacArthur, 1972). Many tropical bird species, at least one-third of the fauna, are considered rare (Lovejoy, 1975; Karr, 1977; Pearson, 1977; Bell, 1982a), and there are very few ecological similarities among them (Pearson, 1980). Rare species tend to use specialized hunting techniques, rely on rare food resources, or utilize large home ranges, sections of which are traversed intermittently. Large home ranges are common among tropical birds, possibly because of scarce or patchily distributed food supplies (Karr, 1976). The fact that Bell (1982a) was able to record over 93% of the lowland rainforest species known regionally in a 2-ha plot over 2 years suggests that at least in New Guinea lowlands the patchiness of distribution is more temporal than spatial. The relatively small number of species involved in New Guinea may be the reason for their more ubiquitous distribution than their Central or South American counterparts. After 225 hours of observation in an area of 15 ha at Maprik in northern New Guinea, Pearson (1975b, 1977) recorded 70% of the regional lowland rainforest avifauna whereas in the lowland rainforest plots of Ecuador, Peru, Gabon, and Borneo he observed smaller percentages of available species in greater numbers of hours. In Australia most rainforest species are distributed widely within their respective ranges; thus within the same region species overlap between the plots is great. It is the small plot size and sporadic appearances of nonrainforest birds that caused reduction in species overlap. It is also possible in some species that the type of specialization is not related to the structure of habitat by which communities are usually classified (Kikkawa, 1982). For example, the mistletoe bird (*Dicaeum hirundinaceum*) in the Australian forest is associated with the mistletoe and not the structure of the forest.

Birds of Disturbed Rainforest and Regrowth Areas

Natural disturbances to rainforest are caused by tree falls. Tree falls are usually local, creating openings in the canopy and encouraging growth of understorey vegetation, and sufficiently frequent to permit maintenance of local tree species diversity through compensatory processes (Connell et al., 1984). Occasionally, extensive areas of rainforest are destroyed by severe cyclones (Webb, 1958) or fire following extreme drought (Beaman et al., 1985). Natural regeneration following extensive damage involves colonization of rainforest trees from sources and the regeneration process may be very slow. In Australia, sclerophyll vegetation (e.g., *Acacia*) appears in the early stages of succession, sometimes mixed with fast-growing fugitive rainforest trees, whereas in New Guinea only the rainforest species are available for colonization (White, 1975).

Natural regeneration following clear felling at Gogol–Naru involved tree species that were a subset of the original flora adapted to the local edaphic conditions. Similarly, the avifauna of regrowth areas was largely derived from bird populations in unlogged forest, though many of the species abundant in intact forests were seldom, if ever, detected in the regrowth. Of the species utilizing the regrowth, some were more abundant in regrowth but most were either equally abundant or more abundant in the original forest. The few migratory species favored regrowth vegetation and several kingfishers responded favorably to conditions created by logging, in particular ponding on level terrain. Three insectivores and two mixed feeders were abundant in the regrowth though virtually absent from the forest areas. They were the only species characteristic of second growth at Gogol–Naru. In other parts of New Guinea (Diamond, 1975; Bell, 1982c) and north Queensland (Kikkawa, 1982), regrowth may contain, in addition, seed eaters and nectar feeders of shrub layers that are normally associated with forest edge. On the whole, there are relatively few species in the New Guinea lowlands that are primarily adapted to a large expanse of secondary growth, although many species are capable of utilizing these areas. Which species form the regrowth subset depends on the local conditions and possibly historical factors of colonization and competition (Diamond, 1970, 1975).

Karr (1976) compared the species composition of bird communities in different stages of forest succession in Africa and in the neotropics, and concluded that shrub and forest faunas in Liberia are less distinct and consist of more habitat generalists than shrub and forest faunas in Panama. He put forward various explanations for the lack of second growth species in Liberia. One is that the African rainforest fauna may be what is left of a fauna that was more diverse and specialized in the past, but now replaced by a set of more opportunistic species following extinctions caused through the isolation of habitat during Pleistocene climatic changes and long periods of human occupation. In contrast, the specialized second growth fauna in the neotropics may have developed in response to extensive river and coastal habitats (Terborgh and Weske, 1969; Lovejoy, 1975). Another explanation is that because of the past scarcity of expansive secondary vegetation there has been little opportunity for a characteristic second growth fauna to develop. While the first explanation seems to hold for Liberia and northern Australia (Kikkawa et al., 1980), the second explanation fits the situation in New Guinea where there is little differentiation of regrowth fauna. It is easier to name species that remain in forest interior and never venture outside or visit regrowth areas than to define the regrowth fauna. They include specialized ground feeders of the dark forest floor and small insectivores of canopy foliage (Bell, 1982c) as well as large frugivores of the canopy and large ground birds such as cassowaries, megapods, and crowned pigeons. In West Malaysia the great majority of bird species are considered to be dependent on the continued presence of undisturbed forest (Medway and Wells, 1971). In the tableland rainforest of north Queensland where endemic species are confined to the forest interior, large-scale disturbances would endanger these species, though minor disturbances caused by selective logging (Severin) and tourism (Lake Eacham) produced only numerical responses (positive at Severin and negative at Lake Eacham) of understorey species. Similar positive responses were observed in selectively logged areas of subtropical montane rainforest in Australia where no invasion of bird species from other habitats occurred (Pattemore and Kikkawa, 1974). In the lowland areas of tropical Australia, natural mosaic vegetation supports a mixed fauna of rainforest and sclerophyll vegetation. Thus a large number of species from sclerophyll and coastal complex habitats are available for colonization in the early stages of rainforest regrowth, diluting the rainforest species that appear in regrowth. There is, however, no distinctive regrowth fauna, and the type of adjacent habitat seems to determine the faunal composition as found at Freshwater Valley and Noah Creek plots. Here opportunistic species are drawn to regrowth areas from both within and outside the rainforest.

Shift in Foraging Heights

Finer vertical subdivision of foraging activity by tropical birds contributes to their higher species diversity in rainforests (MacArthur, 1965; Orians, 1969; Terborgh and Weske, 1969; Karr and Roth, 1971). Additional layering of tropical forests (Terborgh and Weske, 1969) and interspecific competition (Diamond, 1970) are considered to be important factors in producing vertical stratification. Thus the changes of vegetation structure in regrowth are expected to result in lowered stratification of rainforest birds.

For example, *Rhipidura* spp. (fantails) are noted for their vertical separation of foraging activity in the rainforest of New Guinea (Beehler, 1981) and north Queensland (Crome, 1978; Frith, 1984). At Gogol–Naru, *Rhipidura rufiventris* foraged high in the middlestorey of unlogged forest but regularly came much closer to the ground in regrowth vegetation. *R. rufidorsa* generally foraged below *R. rufiventris* in forest areas but was very scarce in regrowth, which may have been due to both increased competition from *R. rufiventris* and changes in habitat structure. In north Queensland *R. rufifrons* forages significantly lower than *R. fuliginosa* in the rainforest of both lowlands (Crome, 1978) and tablelands (Frith, 1984). In the selectively logged plot (Severin) both species increased in number in response to openings in the canopy and increased ground cover, and both were caught frequently in ground nets. Detailed observations by Frith (1984) revealed that these species shifted their foraging heights between seasons, as did other insectivores, in response to fluctuations in insect abundance (Frith and Frith, 1985). They tended to forage lower in the wet season while maintaining the vertical stratification in their preferred heights. Similar differences in foraging heights existed among the nectarivores in New Guinea (Terborgh and Diamond, 1970) and affected their distribution in regrowth areas.

Niches of frugivores are usually defined by the size and type of fruit they utilize (Diamond, 1973; Crome, 1975; Frith et al., 1976) rather than by vertical stratification (Greig-Smith, 1978). However, observations from the canopy platforms at Gogol–Naru revealed that there is vertical stratification among the frugivores utilizing the fruit of emergents and canopy trees. Very few frugivores appeared in the regrowth plots where most of them occupied the crowns of remnant canopy trees consisting of a narrow 10-m band of foliage.

The vertical structuring of the Gogol–Naru birds was similar to that described by Bell (1982b) for the Brown River birds in terms of the proportion of the fauna represented in the catches of ground nets (30–36%). At the regrowth plots in Gogol–Naru 42% of the species recorded were netted, indicating that of the species common to intact forest many were retaining their original vertical ranges in regrowth. At a regrowth plot in Peru (Terborgh and Weske, 1969), 70% of all the species recorded were taken in ground nets, but the maximum height of the vegetation there was only 7 m. In other modified vegetation (coffee and cacao groves), where the canopy was taller and consisted of indigenous trees, Terborgh and Weske netted no more than 50% of the species. They found that scattered trees of the original forest attracted a large contingent of canopy birds, but certain groups (e.g., tanagers and honeycreepers) were better represented than others (e.g., forest ant birds). At Gogol–Naru, large herbivores were poorly represented in the regrowth, whereas most open-air salliers, including migratory species, intensively utilized the remnant tall trees. Pearson (1975a) noted that several southern migrants on a forest plot in Peru sallied from emergent trees. These same species, when present near forest plots in Ecuador and Bolivia, occurred only in open areas or secondary forest. Terborgh and Weske (1969) found that it was the form of the vegetation, and not necessarily the percentage cover of particular layers, that determined the presence of species. Even though the understorey of coffee trees on two of their study plots presented comparable cover to natural undergrowth, it attracted very few species possibly because of scarce food in the uniform, monospecific vegetation. Similarly, the depauperate fauna of the *Eucalyptus* plantation at Gogol–Naru may have resulted from the inability of birds to utilize this exotic vegetation (Driscoll, 1985). The vegetation profile of the plantation also differed considerably to the vertical structure of natural forest and regrowth.

Janzen (1973) cautioned that, even if it is tempting to equate the microenvironments of forest canopy and second growth vegetation, the two actually have little in common in terms of insect faunas, physiognomy, or patterns of flower and fruit production. Pearson (1971) recognized eight species in his plot in Peru as utilizing both forest canopy and the tops of regrowth. However, in commenting on open-area species, he noted that only one species occurred nearly as often in both canopy and low successional vegetation. In Ecuador and Bolivia, there was less tendency than in Peru for birds to specialize in foraging in regrowth as well as across the forest canopy (Pearson, 1975a). Similarly, in Gogol–Naru only three small insectivores seemed to glean readily across the outer layer of vegetation regardless of its height. However, Greenberg (1981) found a considerable overlap of species feeding in low regrowth and the forest canopy on Barro Colorado Island.

Roth (1976) maintained that there was more horizontal and less vertical separation of species in grassland and new regenerating vegetation compared with the later stages of forest succession. Terborgh and Weske (1969) speculated that as tropical vegetation matures, species are able to use particular strata and extend their home ranges within narrow vertical limits. Species in regrowth have fewer options for spatial segregation and must rely on different patches of vegetation rather than different layers for foraging. This pattern of utilization of the regrowth vegetation portrays a patchy mosaic of bird activity that is largely determined by the heterogeneity of habitat structure.

Acknowledgments. Field work in the Gogol–Naru area was made possible by grants from the New Guinea Biological Foundation and the UNESCO Man and Biosphere (MAB) Programme and through the cooperation and hospitality of local village people. The Offices of Forests, Environment and Conservation and the Wildlife Branch of the Department of Natural Resources, Papua New Guinea, assisted in a variety of ways. Field work in Australia was supported by the Australian Biological Resources Study (74/1010) for Cape York Peninsula and the Australian Research Grants Committee (D65/15291) for north Queensland.

References

Beaman, R.S., Beaman, J.H., Marsh, C.W., and Woods, P.V. 1985. Drought and forest fires in Sabah in 1983. Sabah Soc. J. 8, 10–30

Beehler, B. 1981. Ecological structuring of forest bird communities in New Guinea. In: Gressitt, J.L. (ed.), Biogeography and Ecology of New Guinea, Junk, The Hague, pp. 837–861

Beehler, B.M., Pratt, T.K., and Zimmerman, D.A. 1986. Birds of New Guinea, Princeton Univ. Press, Princeton, NJ

Bell, H.L. 1982a. A bird community of lowland rainforest in New Guinea. 1. Composition and density of the avifauna. Emu 82, 24–41

Bell, H.L. 1982b. A bird community of lowland rainforest in New Guinea. 3. Vertical distribution of the avifauna. Emu 82, 143–162

Bell, H.L. 1982c. A bird community of lowland rainforest in New Guinea. 4. Birds of secondary vegetation. Emu 82, 217–224

Bourlière, F. 1983. Animal species diversity in tropical forests. In: Golley, F.B. (ed.), Tropical Rain Forest Ecosystems. A. Structure and Function, Elsevier, Amsterdam, pp. 77–91

Connell, J.F., Tracey, J.G., and Webb, L.J. 1984. Compensatory recruitment, growth, and mortality as factors maintaining rain forest tree diversity. Ecol. Monogr. 54, 141–164

Crome, F.H.J. 1975. The ecology of fruit pigeons in tropical northern Queensland. Aust. Wildl. Res. 2, 155–185

Crome, F.H.J. 1978. Foraging ecology of an assemblage of birds in lowland rainforest in northern Queensland. Aust. J. Ecol. 3, 195–212

Diamond, J.M. 1970. Ecological consequences of island colonisation by southwest Pacific birds. 1. Types of niche shift. Proc. Natl. Acad. Sci. USA 67, 529–536

Diamond, J.M. 1973. Distributional ecology of New Guinea birds. Science 179, 759–769

Diamond, J.M. 1975. Assembly of species communities. In: Cody, M.L., and Diamond, J.M. (eds.), Ecology and Evolution of Communities, Harvard Univ. Press, Cambridge, MA, pp. 345–444

Driscoll, P.V. 1985. The Effects of Logging on Bird Populations in Lowland New Guinea Rainforest. Ph.D. thesis, Univ. of Queensland, Brisbane

Driscoll, P.V. 1986. Activity of understorey birds in relation to natural treefalls and intensive logging in lowland New Guinea rain forest. Proc XVIIIth IUFRO Congress, Div. 1, Vol. I, Ljubljana, Yugoslavia, pp. 620–632

Frith, C.B., and Frith, D.W. 1985. Seasonality of insect abundance in an Australian upland tropical rainforest. Aust. J. Ecol. 10, 237–248

Frith, D.W. 1984. Foraging ecology of birds in an upland tropical rainforest in north Queensland. Aust. Wildl. Res. 11, 325–347

Frith, H.J., Crome, F.H.J., and Wolfe, T.O. 1976. Food of fruit-pigeons in New Guinea. Emu 76, 49–58

Gillison, A.N. 1983. Tropical savannas of Australia and the southwest Pacific. In: Bourlière, F. (ed.), Tropical Savannas, Elsevier, Amsterdam, pp. 183–243

Greenberg, R. 1981. The abundance and seasonality of forest canopy birds on Barro Colorado Island, Panama. Biotropica 13, 241–251

Greig-Smith, P.W. 1978. Social feeding of fruit pigeons in New Guinea. Emu 78, 92–93

Janzen, D.H. 1973 Sweep samples of tropical foliage insects: Effects of season, vegetation types, elevation, time of day, and insularity. Ecology 54, 686–708

Karr, J.R. 1975. Production, energy pathways, and community diversity in forest birds. In: Golley, F.B., and Medina, E. (eds.), Tropical Ecological Systems: Trends in Terrestrial and Aquatic Research, Springer-Verlag, Berlin, pp. 161–176

Karr, J.R. 1976. Within- and between-habitat avian diversity in African and neotropical lowland habitats. Ecol. Monogr. 46, 457–481

Karr, J.R. 1977. Ecological correlates of rarity in a tropical forest bird community. Auk 94, 240–247

Karr, J.R., and Roth, R.R. 1971. Vegetation structure and avian diversity in several New World areas. Am. Nat. 105, 423–435

Kikkawa, J. 1968. Ecological association of bird species and habitats in eastern Australia: Similarity analysis. J. Anim. Ecol. 37, 143–165

Kikkawa, J. 1976. The birds of Cape York Peninsula. Sunbird 7, 25–41, 81–106

Kikkawa, J. 1982. Ecological association of birds and vegetation structure in wet tropical forests of Australia. Aust. J. Ecol. 7, 325–345

Kikkawa, J. 1986. Complexity, diversity and stability. In: Kikkawa, J., and Anderson, D.J. (eds.), Community Ecology, Blackwell, Melbourne, pp. 41–62

Kikkawa, J. 1988. Bird communities of rainforests. Acta XIX Congr. Int. Ornithol., Ottawa, 1986

Kikkawa, J., Lovejoy, T.E., and Humphrey, P.S. 1980. Structural complexity, and species clustering of birds in tropical rainforests. In: Nöhring, R. (ed.), Acta XVII Congr. Int. Ornithol., Deutsche Ornithologen-Gesellschaft, Berlin, pp. 962–967

Kikkawa, J., Monteith, G.B., and Ingram, G. 1981. Cape York Peninsula: Major region of faunal interchange. In: Keast, A. (ed.), Ecological Biogeography of Australia, Junk, The Hague, pp. 1697–1742

Kikkawa, J., and Williams, W.T. 1971. Altitudinal distribution of land birds in New Guinea. Search 2, 64–65

Kwapena, K. 1985. Tropical rain forests and plantation forestry in Papua New Guinea. In: Davidson, J., Pong, T.Y., and Bijleveld, M. (eds.), The Future of Tropical Rain Forests in South East Asia, Commission on Ecology Papers No 10, pp. 91–100, IUCN, Gland, Switzerland

Lamb, D. 1977. Conservation and management of tropical rainforest: A dilemma of development in Papua New Guinea. Env. Conserv. 4, 121–129

Laverack, P.S., and Stanton, J.P. 1977. Vegetation of the Jardine River catchment and adjacent coastal areas. Proc. Roy. Soc. Qd 88, 39–48

Lovejoy, T.E. 1975. Bird diversity and abundance in Amazon forest communities. Living Bird 13, 127–191

MacArthur, R.H. 1965. Patterns of species diversity. Biol. Rev. 40, 510–533

MacArthur, R.H. 1972. Geographical Ecology: Patterns in the Distribution of Species. Harper and Row, New York

MacArthur, R.H., Recher, H., and Cody, M.L. 1966. On the relation between habitat selection and species diversity. Am. Nat. 100, 319–332

McClure, H.E. 1967. The composition of mixed species flocks in lowland and sub-montane forests of Malaya. Wilson Bull. 79, 131–154

McIntosh, D.H. 1974. Progress Report for the 10th Commonwealth Forestry Conference, Department of Forests, Port Moresby

Medway, Lord, and Wells, D.R. 1971. Diversity and density of birds and mammals at Kuala Lompat, Pahang. Malay. Nat. J. 24, 238–247

Meijer, W. 1970. Regeneration of tropical lowland forest in Sabah, Malaysia forty years after logging. Malays. For. 33, 204–229

Morisita, M. 1971. Composition of Iw-index. Res. Popul. Ecol. 13, 1–27

Orians, G.H. 1969. On the evolution of mating systems in birds and mammals. Am. Nat. 103, 589–603

Pattemore, V., and Kikkawa, J. 1974. Comparison of bird populations in logged and unlogged rain forest in Wiangarie State Forest, N.S.W. Aus. Forestry 37, 188–198

Pearson, D.L. 1971. Vertical stratification of birds in a tropical dry forest. Condor 73, 46–55

Pearson, D.L. 1975a. The relation of foliage complexity to ecological diversity of three Amazonian bird communities. Condor 77, 453–466

Pearson, D.L. 1975b. Survey of the birds of a lowland forest plot in the East Sepik District, Papua New Guinea. Emu 75, 175–177

Pearson, D.L. 1977. A pantropical comparison of bird community structure on six lowland forest sites. Condor 79, 232–244

Pearson, D.L. 1980. Patterns of foraging ecology for common and rarer bird species in tropical lowland forest communities. In: Nöhring, R. (ed.), Acta XVII Congr. Int. Ornithol., Deutsche Ornithologen-Gesellschaft, Berlin, pp. 974–978

Pedley, L., and Isbell, R.F. 1971. Plant communities of Cape York Peninsula. Proc. Roy. Soc. Qd. 82, 51–74

Rand, A.L., and Gilliard, E.T. 1967. Handbook of New Guinea Birds, Weidenfeld and Nicolson, London

Roth, R.R. 1976. Spatial heterogeneity and bird species diversity. Ecology 57, 773–782

Terborgh, J., and Diamond, J.M. 1970. Niche overlap in feeding assemblages of New Guinea birds. Wilson Bull. 82, 29–52

Terborgh, J., and Weske, J.S. 1969. Colonisation of secondary habitats by Peruvian birds. Ecology 50, 765–782

Tracey, J.G. 1982. The Vegetation of the Humid Tropical Region of North Queensland. CSIRO, Melbourne

Tracey, J.G., and Webb, L.J. 1975. Key to the Vegetation of Humid Tropical Region of North Queensland, plus 15 maps at 1:100,000 scale. CSIRO Div. Plant Industry, Canberra

Webb, L.J. 1958. Cyclones as an ecological factor in tropical lowland rainforest, North Queensland. Aust. J. Bot. 6, 220–228

Webb, L.J. 1966. The identification and conservation of habitat-types in the wet tropical lowlands of North Queensland. Proc. Roy. Soc. Qd. 78, 59–86

Webb, L.J. 1968. Environmental relationships of the structural types of Australian rainforest vegetation. Ecology 49, 296–311

Webb, L.J., and Tracey, J.G. 1981a. Australian rainforests: patterns and change. In: Keast, A. (ed.), Ecological Biogeography of Australia, Junk, The Haque, pp. 607–694

Webb, L.J., and Tracey, J.G. 1981b. The rainforests of northern Australia. In: Groves, R.H. (ed.), Australian Vegetation, Cambridge Univ. Press, Cambridge, UK, pp. 67–101

Webb, L.J., Tracey, J.G., and Williams, W.T. 1976. The value of structural features in tropical forest typology. Aust. J. Ecol. 1, 3–28

White, K.J. 1971. The lowland rain forest in Papua New Guinea. Pacific Science Association Pre Congress Conference, Bogor, 12–17 August 1971

White, K.J. 1975. The effect of natural phenomena on the forest environment. Presidential Address to the PNG Sci. Soc., 26 March 1975, Department of Forests, Port Moresby

Whyte, I.N. 1975. Land classification and mapping for reforestation planning in Papua New Guinea. Unpublished manuscript, Papua New Guinea Forestry College, Bulolo

Willson, M.F., and Moriarty, D.J. 1976. Bird species diversity in forest understorey: Analysis of mist net samples. Oecologia 25, 373–379

Willson, M.J., Anderson, S.H., and Murray, B.G. 1973. Tropical and temperate bird species diversity: Within-habitat and between-habitat comparisons. Caribb. J. Sci. 13, 81–90

Winter, J.W., Bell, F.C., Pahl, L.I., and Atherton, R.G. 1987. Rainforest clearfelling in northeastern Australia. Proc. Roy. Soc. Qd. 98, 41–57

6. Mammalian Species Richness in Tropical Rainforests

François Bourlière

The mammal fauna of tropical rainforests is generally considered as being reasonably well known, though a few species of bats, rodents, and other small mammals are still discovered every year, and the taxonomic status of many other forms remains unsettled. However, very little is known of the ecology and behavior of most rainforest species as compared to their relatives living in more open savannas. As for the structure and dynamics of rainforest mammalian communities, its study has hardly begun, though their species richness is generally emphasized. Is this last viewpoint justified? This is what I would like to consider in reviewing the scanty data available. The possible determinants of species diversity among tropical mammals will also be briefly discussed, and compared with those considered as important in enhancing and maintaining species diversity in other vertebrates.

Limitations of the Available Data

Before considering the data at hand, it seems appropriate to call the reader's attention on some of the conditions that make the study of tropical rainforest mammal communities so difficult. First of all, and contrary to what is the case in more open savannas, mammals are not easy to observe in the "closed" environment of a tropical rainforest. This has been noticed since the days of the early naturalist explorers and cannot be attributed only to recent human interference or increased hunting pressure. Besides the troops of monkeys that may sometimes be seen or heard in the treetops, or an occasional squirrel running on a branch, the observer proceeding as silently as possible into a rainforest scarcely sees any mammal by day. The situation is different during the night; bats are seen or heard flying about everywhere, and the beam of a torch often reveals pairs of eyes glowing in the dark, peering at the intruder from high into the trees or from the undergrowth. Whether it be by day or by night, larger ungulates or carnivores are seldom flushed, generally singly or in small family parties. Therefore, observation is in most cases of limited value for the mammalogist in a rainforest environment; but this does not mean that it is useless. As a matter of fact, it is the only method available for the study of larger arboreal species such as monkeys and apes, and a few other diurnal mammals. But it is a time-consuming procedure that must be carried out by well-trained observers (see, for instance, N.R.C., 1981, for primates). For many shy diurnal and nocturnal species, it will be necessary to rely on contacts along line transects, or on indirect methods such as tracking or collecting fecal pellets, to determine the presence and evaluate the relative abundance of large terrestrial species.

The trapping of small ground-living or scansorial mammals is equally difficult in tropical rainforests. All kinds of traps and baits are selective, all the more so when dietary resources are abundant and varied, and the "enclosed quadrat" technique, with hand digging and opening of all holes and burrows (Bellier, 1967; Dieterlen, 1967), is difficult to apply to the forest floor. As for the use of mist nets for the sampling of the bat fauna, it is also selective, and only allows the capture of species flying low in the understorey, along forest paths or creeks; the canopy fauna remains out of reach. The systematic exploration of all possible roosts remains the only way to get hold of some species.

Another difficulty arises when one tries to assign limits to a rainforest mammal community: what spatial scale to use, small or large? A tropical rainforest, whether primeval or secondary, never is a homogeneous environment. It is a landscape made of a mosaic of habitats, representing various seral stages, or corresponding to tree communities peculiar to certain edaphic conditions. Habitat heterogeneity is the rule, each habitat harboring eventually its own assemblage of small mammal species; only the larger, wide-ranging, or more generalist species can make use of a rainforest landscape as a whole.

The limitations just mentioned must be kept in mind when interpreting the figures given in Table 6.1. These figures, partly taken from the literature and partly based on personal communications, refer to the species known to live (or that have lived) in a number of tropical areas adequately studied, and generally protected against major human interferences for a number of years. Besides information (when available) on mean annual rainfall, average altitude, and surface studied, the table gives the total number of mammal species in each rainforest area, the number of bats being indicated in parentheses following the total. Next to rainforest figures, similar information on more "open" tropical environments is given for comparative purposes. One will immediately notice that the size of the areas surveyed varies considerably, from a few square kilometers to a few hundred, and even thousands in the case of African savannas. As the number of species, as well as the heterogeneity of the habitat, increases with the size of the sampling area, it is fair to compare sites of similar size only.

Overall Species Richness

These limitations being taken into account, several conclusions can be drawn from the figures in Table 6.1. First, most tropical landscapes generally, but not always, harbor more species of mammals than homologous temperate landscapes of similar size. For instance, only 58 species of native mammals live in the famous Bialowieza forest, eastern Poland, 13 of which are bats (Z. Pucek, personal communication). Second, the number of mammal species is often as large in African savannas and forest-savanna mosaic as in rainforest areas. (Compare, for instance, the figures of the Garamba, Serengeti, and Kruger National Parks, and also those of Lamto, with the figure for Makokou.) Third, the number of species of mammals is always smaller on "miniconti-nents" such as Madagascar (594,000 km²) and New Guinea (800,000 km²) than in comparable landscapes on large continents. On large, isolated continental islands mammals become extremely scarce; there are only five species of native mammals, all bats, on New Caledonia (19,000 km²). Fourth, terrestrial mammals are more numerous in species on the African mainland than anywhere else in the tropics. Conversely, bats make up to 59% of the number of mammal species present on some tropical American study sites. Finally, the sites located on former postulated Pleistocene forest refuges, such as Cocha Cashu or Makokou, seem more species-rich than those situated outside of these former "forest islands."

Structure of Rainforest Mammalian Communities

The various orders of rainforest mammals differ widely in form, function, and behavior. Their ways of life are therefore very dissimilar, and the mere number of species found at a given site does not mean much about the structure and functioning of the mammalian community.

No wonder, then, that the first detailed studies on the community structure of rain-forest mammals were carried out on those taxonomic groups that were the most intensively studied during the past two decades: monkeys and apes (Bourlière, 1985; Richard, 1985), bats (Fleming et al., 1972; Bonaccorso, 1979; LaVal and Fitch, 1977), shrews (Brosset, 1988), and rodents (Payne, 1979; Emmons, 1980; Whitten, 1981; Duplantier, 1983). A most interesting attempt to study the ecological relationships between all plant-eating mammals of North-East Gabon has been made by Emmons et al. (1983).* Unfortunately, nothing similar has been attempted so far in the neotropics or South East Asia.

The preliminary generalizations that follow will therefore be limited to the most obvious characteristics of rainforest mammalian communities. Further studies will have to validate them and to describe the finer mechanisms of ecological partitioning among sympatric (and sometimes syntopic) species.

---

*When the present chapter was already in press, a broad comparison of the mammal faunas of north-east Gabon and French Guiana was also published by Dubost (1987, issued in February 1988).

F. Bourlière

Table 6.1. Tropical Species Richness: Native Mammals

Study Sites	Mean Annual Rainfall (mm)	Altitude (m)	Surface Studied (km²)	Number of Species[a]	Sources
Rainforests					
La Selva Reserve, Costa Rica	3991	35–150	7.3	137 (91)	1
Barro Colorado, Panama	2600	25–165	15.0	96 (46)	2
Fort Sherman, Cristobal, Panama	3000	5		70 (31)	3
Rio Alto, eastern Peru	ca2900	170–790		109 (38)	4
Rio Curanja, eastern Peru	ca2500	300		122 (57)	4
Cocha Cashu, eastern Peru	2000	380	75.0	99 (25)	5
Parque Nacionale Iguazu, Argentina	1700	200–300	550	52 (7)	6
Bukit Lanjan Forest Reserve, W. Malaysia	ca2300	40	6.5	96 (40)	7
Makokou, Gabon, whole area	ca2000	470–925	ca17000	126 (34)	8
Makokou, Gabon, M'Passa reserve	1730	500	100	114 (33)	8
Mount Kaindi, Papua New Guinea	ca2000	1600–1900		50 (11)	9
Baiyer River Sanctuary, Papua New Guinea	2500	1200	7.4	22 (7)	10
Analamazoatra (Perinet), Madagascar	1700	900	8.0	38 (9)	11
Forest-Savanna Mosaic					
Lamto, Ivory Coast, IBP study site	1300	70–110	25	116 (32)	12

Savannas and Woodlands

Palo Verde Refuge, Costa Rica	1000–1500	100–200	475	133 (85)	13
Acurizal Ranch, Brazil	1120	100	70	64 (21)	14
Serra do Roncador, Brazil	1370	90–260	ca400	63 (27)	15
Masaguaral Ranch, Venezuela	1462	63	34	71 (42)	16
Município de Exu, Pernambuco, Brazil	400–1000	400–500	ca200	52 (31)	17
Jabiluka, N.T., Australia	1200–1500	90–260	ca400	63 (27)	18
Garamba National Park, Zaire	1500	785	4800	138 (38)	19
Serengeti National Park, Tanzania	500–1000	1200–1700	13250	130 (25)	20
Kruger National Park, Transvaal	375–750	165–660	19485	141 (41)	21
Hluhluwe-Umfolozi Game Reserve, Natal	780	60–650	920	87 (13)	22
Nylsvley Nature Reserve, Transvaal	694	1080–1100	7.45	68 (7)	23
Fété-Olé, Senegal	319	40	100	36 (5)	24

[a]Number of bats indicated in parentheses following the total.

Sources: (1) Wilson, 1983; (2) Koford et al., 1977; (3) Fleming, 1973; (4) Patton et al., 1982; (5) Terborgh et al., 1984; (6) Crespo, 1982; (7) Harrison, 1959; Lim Boo Liat et al., 1974; (8) ECOTROP, 1979, updated by A. Brosset, J.M. Duplantier & A. Gautier-Hion, pers. comm.; (9) Gressitt & Nadkarni, 1978; (10) R.D. MacKay, pers. comm.; (11) Eisenberg & Gould, 1970, updated by A. Albignac, M.E. Nicholl & R.L. Peterson, pers. comm.; (12) Bourlière et al., 1974; Bellier, 1959, 1974; Thomas, 1982; Heim de Balsac, 1974, updated by V. Aellen & F. de Bree; (13) Wilson, 1983; (14) Schaller, 1983; (15) Pine et al., 1970, updated; (16) Eisenberg et al., 1979; (17) Mares et al., 1981; (18) Kerle & Burgman, 1984; (19) Verschuren, pers. comm.; (20) Sinclair & Verschuren, pers. comm.; (21) Piennaar, 1963, 1964; Rautenbach et al., 1985; (22) Bourquin et al., 1971; (23) Jacobsen, 1977; (24) Poulet, 1972.

Activity Rhythm

Most rainforest mammals are definitely nocturnal in habits. At Barro Colorado, Panama, 83 (85.5%) out of the 97 species listed by Koford et al. (1977) are wholly or predominantly nocturnal. Only four species of monkeys, the two squirrels, one sloth, one anteater, the coati, and the agouti are active by day, and four other species forage by day as well as by night. In the Cocha Cashu study area, Peruvian Amazonia, 55 (55.5%) out of the 99 species recorded by Terborgh et al. (1984) are nocturnal, and only 22 diurnal, the others being active both day and night. Here again, the typically diurnal species are monkeys, squirrels, the agouti, and the acouchi, two carnivores and two artiodactyls. In the Makokou area, North-East Gabon, 100 (79.4%) out of the 126 species are nocturnal. Once more, the 11 species of monkeys and apes, and the nine squirrels make up the bulk of diurnal mammals (Charles-Dominique, 1975; Emmons et al., 1983). Within the mountain forest (1500–1900 m) of Mount Kaindi, Papua New Guinea, 41 (82%) species out of the 50 reported to be present by Gressitt and Nadkarni (1978) are crepuscular or nocturnal, the nine others being reported as being active day and night by Ziegler (1982). The percentages given by Davis (1962) for the lowland rainforest mammals of North Borneo are rather similar: only 30% of the 77 species are diurnal, 48% are nocturnal, and 22% active by day and by night. Here again, the diurnal component is accounted for entirely by treeshrews, primates, and squirrels. This nocturnal activity of most rainforest mammals strongly contrasts with the predominantly diurnal habits of birds living in the same environments (Charles-Dominique, 1975).

Foraging Zones

Mammals may be found everywhere within a tropical rainforest. Only fossorial species that live permanently in the ground are entirely missing from this environment, though some Old World shrews and a strange Papuan rodent (*Mayermys*) can spend most of their life foraging into the litter. Amphibious riverine mammals are also rare, but present in all rainforest areas. However, most of them (water oppossums, water shrews, otter shrews, water rats, and otters) very seldom wander far from the waterside; only the pygmy hippopotamus from Western Africa sometimes venture outside the riverine forest. Most of the rainforest mammals live on dry land, exploiting all the food sources of their tridimensional environment. Until we know better the various parameters of their habitat niches, we must satisfy ourselves with a rather simplistic classification based on their preferential use of one of the four major forest levels: ground level (terrestrial species), understorey (scansorial or terrestrial-arboreal species), middle and canopy levels (arboreal species), and the space between and above the trees (aerial species). As it is often difficult to draw a sharp boundary between the scansorial and arboreal foraging zones, and as different authors often use different criteria to do so, the sum total of the scansorial and arboreal species has also been added to Table 6.2.

It seems obvious, on the basis of the few figures given in Table 6.2, that the occupancy of the various forest levels is very different in the Old World and in the New World. Terrestrial mammals are more numerous in the tropical rainforests of the former, whereas there are more species of bats in the neotropics. On the other hand, scansorial and arboreal species of mammals represent a very similar percentage of the rainforest mammalian communities in the Western and Eastern hemispheres.

Table 6.2. Percentages of Species Using the Four Major Foraging Zones, at Four Rainforest Sites (Round Figures)

	Foraging Zones					
	I	II	III	Total of	IV	
Sites	Terrestrial	Scansorial	Arboreal	II & III	Aerial	Source
Cristobal, Panama	24	20	13	33	43	1
Barro Colorado, Panama	25	11	17	28	47	2
Makokou, Gabon	42	12	20	32	26	3
Mount Kaindi, Papua New Guinea	42	12	24	36	22	4

Sources: (1) Fleming, 1973; (2) Eisenberg & Thorington, 1973; Koford et al., 1977; (3) ECOTROP, 1979; Emmons et al., 1983; (4) Species list after Gressit & Nadkarni, 1978; foraging layers after Ziegler, 1982.

Diets

The food habits of rainforest mammals are very unevenly known. The diets of some taxonomic groups, such as bats, primates, ungulates, and rodents, have been more often investigated than those of most marsupials, insectivores, and small carnivores. Pending quantitative studies of the food preferences of such groups, including their seasonal variations, present day ecologists are compelled to rely on qualitative, if not anecdotal, information. For the time being, therefore it is only possible to get an approximate view of the partitioning of food resources among rainforest mammals.

To make things worse, the definitions of the various food categories given by different authors frequently vary, and the diet of many mammals too often falls "in between" the categories most commonly used. To organize the available data and look for patterns, only two possibilities remain: (1) to distinguish only two broad categories, "vegetarians" (or primary consumers) and "carnivores" (or secondary consumers), as in Table 6.3; or (2) to split these two categories into more specific diets, such as frugivores, folivores, granivores, insectivores, and carnivores. However, mixed diets are frequent, such as frugivores-folivores or frugivores-insectivores. To avoid too great a number of feeding categories, it is generally agreed, following Emmons et al. (1983), to include mixed feeders into the category that represents more than 60% of their diet. Thus, frugivores-insectivores and granivores have been pooled with frugivores, and herbivores with folivores, in Table 6.4.

Table 6.3. Percentages of Primary and Secondary Consumers in Four Rainforest Mammal Communities (Round Figures)

Sites	Primary Consumers	Secondary Consumers	Sources
Cristobal, Panama	63	37	1
Barro Colorado, Panama	59	41	2
Makokou, Gabon	53	47	3
Mount Kaindi, Papua New Guinea	56	44	4

Sources: (1) Fleming, 1973; (2) Eisenberg & Thorington, 1973; Koford et al., 1977; (3) ECOTROP, 1979; Emmons et al., 1983; (4) Species list after Gressitt & Nadkarni, 1978.

Table 6.4. Percentages of the Major Types of Diet in Four Rainforest Mammal Communities (Round Figures)

Sites	Feeding Habits				
	Frugivores	Folivores	"Insectivores" (Invertebrate Eaters)	Carnivores	Source
Cristobal, Panama	54	9	20	17	1
Barro Colorado, Panama	53	6	19	22	2
Makokou, Gabon	44[a]	9	38	9	3
Mount Kaindi, Papua New Guinea	32[b]	24	42	1	4

[a] Including one species classified as frugivorous-folivorous by Emmons et al. (1983).
[b] Nectarivores, pollen, and sap eaters added to the frugivores.
Sources: (1) Fleming, 1973; (2) Eisenberg & Thorington, 1973; Koford et al., 1977; (3) ECOTROP, 1979; Emmons et al., 1983; (4) diets based upon Menzies & Dennis (1979), Ziegler (1982) and Nowak & Paradiso (1983).

Some interesting patterns emerge from these tables. First, the percentages of primary and secondary consumers are very similar at the four sites concerned (Table 6.3). Second, frugivorous and carnivorous species are more numerous at the two neotropical sites, whereas there are more insectivorous mammals at both Makokou and Mount Kaindi. Third, the percentages of carnivores is much smaller in Gabon and New Guinea than in Panama. Fourth, the percentage of folivores is particularly high at Mount Kaindi, whereas there are less frugivorous mammals here than elsewhere (Table 6.4).

These differences between sites are not random events. The abundance of frugivorous and carnivorous species in Panama is obviously due to the larger number of neotropical bats with such feeding habits, as opposed to the situation prevailing in Gabon and New Guinea. Finally, the high number of folivores at Mount Kaindi is mostly ascribable to the fact that shoots, fresh leaves, and nectar, rather than fruit, make up the bulk of the vegetable diet of most Papuan marsupials. The larger percentage of insectivores in the Mount Kaindi mammalian community is quite likely due to the small size of most secondary consumers in the rainforests of Papua New Guinea.

Relative Abundance of Species

It has long been pointed out that most of the plant and animal species are scarce in tropical rainforests. This applies to mammals (Fleming, 1973) as well as to the other classes of vertebrates. Among the 99 species of mammals listed by Terborgh et al. (1984) from Cocha Cashu, 58 are considered to be uncommon or rare. The situation is similar at Makokou. Brosset (1966) notes that one species of shrew (*Crocidura poensis*) is represented by as many individuals in the catches as the nine other sympatric shrew species taken together. Among bats, *Hipposideros caffer* makes up more than 95% of all the captured individuals belonging to 34 different species. As for the myomorph rodents, *Hylomys stella* is by far the most numerous murid taken at all times of the year on the forest floor (Duplantier, 1982). Among the squirrels, two species are definitely more often seen than the six others (Emmons, 1980). Obviously, part of this apparent scarcity may be due to observational or trapping bias, but these cannot provide the whole answer.

Determinants of Species Richness

There is no need to review again the various hypotheses put forward to explain the increased species diversity in the tropics. I will confine myself here to a few remarks more particularly relevant to the case of mammals.

There are three different categories of factors that may contribute to a high species richness in a given area: historical and physical factors on the one hand, and biotic factors on the other. The first two categories are intimately linked; they can explain why some tropical areas are more species-rich than others, and why the taxonomic composition of the mammal community differs between areas otherwise very similar in physiography and climate. On the other hand, biotic factors should be able to tell how such species-rich assemblages can last for a long time, instead of evolving toward simpler communities with a smaller number of species and larger populations.

Historical Factors

The striking differences in species richness and taxonomic structure between South America and Africa, continental Africa and Madagascar, or Southeast Asia and New Guinea can only be understood if we consider their geological history since the late Cretaceous, and the environmental changes which took place in the meantime. The Pleistocene glaciations must not be forgotten, as they had an important impact on the extent of rainforest areas in the lower latitudes. We have already alluded to the high species richness and number of endemics of some of the postulated forest refuges in South America and Africa. It is only in these areas, where climatic conditions remained stable for millions of years, that the "time hypothesis" can hold to be true.

Physical Factors

Geological history is also responsible for some of the physical parameters that influence species richness in mammals, the size of the land masses, their isolation, and their climate. The importance of the first two factors is particularly obvious in the case of islands. In Melanesia, for instance, New Britain, which is close to New Guinea, harbors 32 species of mammals out of which 23 are bats (Koopman, 1979), whereas New Caledonia, which is much farther, has only 5 species of native mammals, all bats.

The third physical factor that probably played an important role in fostering high mammalian species richness in tropical rainforests in the evenness of climate, particularly temperature. One cannot but be struck by the number of small and medium-sized rainforest mammals that are poor thermoregulators (marsupials, insectivores, bats, prosimians, edentates) and could not possibly withstand highly seasonal climates, particularly low winter temperatures.

Biotic Factors

A number of ecological mechanisms contribute to the maintenance of high mammalian species diversity in tropical rainforests. They are not mutually exclusive and their relative importance certainly varies from one taxonomic group to another, and also in different kinds of rainforest communities.

Plant Species Diversity

It is commonly felt that diversity begets diversity and that the species richness of primary consumers increases with that of the plants on which they feed. But this relationship is certainly not a simple one. Floristic richness is probably more important for phytophagous insects than for plant-eating mammals. Although some examples of food specialists may be found in most mammalian orders and families, "generalists" are apparently much more numerous. The variety of plant life forms and of "architectural models" (Hallé and Oldeman, 1970) in rainforest vegetation is probably more important than floristic richness per se. They have a direct bearing on the physical structure of the forest itself, providing a variety of supports for locomotion, and of breeding and sleeping sites. Emmons and Gentry (1985) found that different forest structures in different parts of the tropics were correlated with different modes of locomotion in arboreal vertebrates.

Sustained High Primary Production

Sustained high primary production in tropical rainforests has long been considered by itself to be a factor favoring high species richness among animals. This is difficult to prove in our case; only a small part of the primary production is usable by plant-eating mammals, and the precise nutritional requirements of most mammal species remain little known. For some monkey species it has been shown that only a small fraction of the fruits available is actually eaten (Sourd and Gautier-Hion, 1986). If many fruits, a number of seeds, nectar, and gums are favorite items, and even staple food, for many mammals, they are also eaten by a much larger number of vertebrate and invertebrate consumers. As for the leaves, they can only be eaten by mammals when they are fresh, poor in tanins and other "secondary subtances." The sustained character of the primary productivity is probably more important than its actual amount. Although true "aseasonal" rainforests are rare, the seasonal food shortages that occur in the humid tropics (see, for instance, Leigh et al., 1982) are always far less serious than the dry season shortages in tropical savannas or the winter food shortages in temperate latitudes. However, that soil fertility and undergrowth density might be positively correlated with species richness and density of nonflying rainforest mammals is suggested by the findings of Emmons (1984) in Amazonia; when comparing densities and number of species at seven localities in Peru, Ecuador, and Brazil, she found that small mammal groups that showed large reductions in numbers of individuals also had fewer species on nutrient-poor soils.

Environmental Heterogeneity

The environmental heterogeneity of tropical rainforests quite likely also play a role in furthering species richness, as it also contributes to increase the number of microhabitats available for smaller mammals. This three-dimensional mosaic of spatial niches has little counterpart in more open landscapes, or even in temperate forests;

it also increases the variety of food resources, as those available in the various stages of regeneration after tree falls, for example, differ from those of the patches of mature rainforest.

Sedentariness and Restricted Range

The sedentariness and restricted range of most rainforest mammals is also a factor to be taken into account. Most of the smaller species have very restricted home ranges, and their populations are small, as already mentioned. In many cases sedentariness can be related to the patchy distribution of specific food sources, or habitat characteristics. Conversely, rainforest ungulates, or terrestrial primates like mandrills, cannot find enough fallen fruits or browse on a small area and have to nomadize on larger home ranges. Such mammal groups are less species-rich than the smaller rodents.

Reduced Interspecific Competition

Most ecology textbooks mention reduced interspecific competition as a possible cause for high species richness in the humid tropics. Such a result could be achieved in two different ways. Some consider that a tropical rainforest is a "saturated" community into which a large number of narrowly specialized species are packed, whose small niches do not overlap much because they are food and/or habitat specialists. This results in a lesser interspecific competition. Other theoretical ecologists feel that coexistence is made possible because most of the co-occurring species are kept at low density levels through a heavy predation or parasitic pressure.

Unfortunately, so few niche parameters of tropical rainforest mammals have been properly studied so far that it is hard to validate or disprove these hypotheses, though the large number of species sharing similar life-styles within rainforest communities (the "guilds") does not favor a prominence of "specialists" in tropical rainforests. However, it must always be kept in mind that a human observer is often unable to perceive subtle differences of diet or foraging habits that may be of the utmost importance for the species observed.

Another argument in favor of a reduced interspecific competition among sympatric rainforest mammals is the existence of polyspecific foraging associations. The best example is that of the African and South American rainforest monkeys (Gautier and Gautier-Hion, 1969; Struhsaker, 1981; Gautier-Hion et al., 1983; Terborgh, 1983). Up to three species of similar diet and body size, often closely related taxonomically, group together for most of the day to exploit the same kind of resources, and this association can last for months. Its major function seems to be to improve foraging efficiency and predator avoidance. It is worthwhile to note here that such polyspecific associations of monkeys are far more frequent and numerous in participants in tropical rainforests than in tropical savannas.

Impact of Mammalian Species Richness on Ecosystem Function

Vertebrates — birds and mammals particularly — exert a very important feedback control on the vegetation and play a significant role in maintaining the homeostasis of the forest ecosystem. The pollination of many rainforest trees and shrubs by frugivorous and nectarivorous bats has been known for a long time, but the similar role of small marsupials and prosimians has been discovered much more recently (Sussman and Raven, 1978).

But frugivorous mammals play a much more important role in ensuring seed dispersal in tropical rainforests. Here, a partitioning of roles between birds and mammals is apparent. It depends both on the body size of the animals and on their locomotor abilities. Volant animals, such as birds and bats, or large itinerant mammals, such as elephants, can disperse seeds much further from the parent trees than scansorial mammals (such as rodents or primates) or ground-living sedentary species (such as agoutis, peccaries, and forest duikers). A good example of the importance of zoochory in a tropical community has been provided by Thomas (1982). The "seed rain" resulting from the consumption of fruit by a community of fruit bats at Lamto has been measured at the edge of a gallery forest. The results were astonishing: bats accounted for 98.2% of all the seeds deposited at sites potentially colonized by these plant species. Seeds found in bat feces germinated more quickly and achieved a higher germination success than those removed from fresh fruits. De Foresta et al. (1984) also reported an important nocturnal seed rain by bats at the edge of a rainforest in French Guiana.

Conversely, large terrestrial mammals can also exert an important predation pressure on the early growth stages of forest trees and shrubs. This is not often the case at the present time, as "game" is heavily hunted almost everywhere. But the situation is different in some remote areas. In a piece of "pristine" rainforest of French Guiana, Charles-Dominique (personal communication) was recently impressed by the damage done by the tapirs, deer, peccaries, and agoutis to seedlings, saplings, and young shoots. When undisturbed and numerous enough, these mammals unquestionably contribute to keep the undergrowth more "open" than in similar forests subject to a heavy hunting pressure.

Conclusion

The high species richness of rainforest mammals is a well established fact in most parts of the world, though its causes still remain obscure. However, I would like to draw attention to two points. First, the species richness of mammals in tropical rainforests is always lower than that of the sympatric bird community, which can be as much as five times as numerous in species (at Cocha Cashu, for instance). It only exceeds that of the sympatric reptiles (Table 6.5). The reason for such a discrepancy between the two classes of warm-blooded vertebrates is not obvious at once. However, a closer study of the feeding guilds of rainforest birds and mammals quickly reveals a basic difference between the two communities. There are many more "insectivores" (exclusive and partial) among rainforest birds than among sympatric mammals: up to 75% in North-East Gabon (Erard, 1986); 65% in Peruvian Amazonia (Terborgh, in Beehler, 1981); 55% in French Guiana (Erard, 1986); and 52% in Papua New Guinea (Beehler, 1982) – against 42 to 19% only among mammals (Table 6.4). Because they are able to feed on small mobile invertebrate prey at a low energetic cost in a tridimensional environment, birds have a definite advantage over scansorial and arboreal mammals that remain dependent on plant supports (branches and twigs of adequate size and inclination), along which

Table 6.5. Comparative Species Richness of Vertebrates (Amphibians Excluded) in Seven Rainforest Sites

Sites	Birds[a]	Mammals[b]	Reptiles[c]
La Selva Reserve, Costa Rica	385 (96)	137 (91)	74 (23)
Cocha Cashu, eastern Peru	536 (22)	99 (25)	—
Barro Colorado, Panama	366 (83)	97 (46)	64 (18)
Makokou (M'Passa), N.E. Gabon	363 (52)	114 (33)	65 (15)
Gogol Forest, Papua New Guinea	162 (?)	27 (?)	34 (?)
Baiyer River, Papua New Guinea	163 (15)	22 (7)	20 (9)
Perinet, Madagascar	73 (2)	38 (9)	43 (26)

[a] Number of long-range migrants indicated in parentheses following the total.
[b] Number of bats indicated in parentheses following the total.
[c] Number of lizards indicated in parentheses following the total.

they move at a higher energetic cost. Among mammals, bats enjoy the same advantage than birds and consequently are able to make up a large, if not the largest, part of the mammalian rainforest community. Fleming (1973) also noticed that a large part of the increase of mammalian species richness in the tropics can be ascribed to the increase in number of bat species.

Generally speaking, it appears that species richness is fostered in those vertebrate groups whose locomotor abilities allow access, at low energy cost, to food categories out of reach of species confined to terrestrial or arboreal substrates. Hence, the high species richness of birds, as compared to mammals and reptiles.

On the other hand, the enforced sedentary life-style of small populations of small terrestrial mammals in patchy rainforest habitats might also offer some evolutionary advantages. Gene flow between local populations being all the more reduced as the distances between patches increase, such populations will tend to differ more and more from each other, thus increasing the rate of speciation. This might also help to explain why there are so many genera of small rainforest terrestrial mammals represented by several species in a same area.

In brief, high species richness is not a general attribute to all mammalian orders within tropical rainforests. Only those taxonomic groups able to exploit at low energetic cost the most abundant plant and animal food sources (fruit, nectar, fresh leaves, and invertebrate prey) in this tridimensional environment show high species diversity. This is the case for all volant mammals (bats), and for most small and medium-sized arboreal mammals, from the smallest marsupials to the smaller monkeys. In contrast, larger mammals confined to the forest floor where food is always scarce are much less numerous in species. That locomotor abilities can and do play a central role in foraging space partitioning has recently been demonstrated by McKenzie and Rolfe (1986) in their study of bat communities in the northwest Australian mangroves. Differences in species aerodynamic characteristics were related to vertical and horizontal foraging microhabitats.

As for the relative weight to be given to the various determinants of community structure in rainforest mammals, it is obviously too early to draw any definite conclusion. All that can be said is that although historical factors and chance events certainly play a more or less important role in the shaping of most communities, the data at hand do not support the view that species assemblages of rainforest mammals are structured by chance alone. Biotic interactions quite likely also play an important role in patterning the ecological and morphological relationships of mammal species in tropical rainforests. For instance, the way the weights of the 66 species of mammalian primary consumers of a North-Eastern Gabon forest are nearly quite uniformly distributed over a range of five orders of magnitude is quite remarkable (Emmons et al., 1983, fig. 1). Similarly, the sizes of the 11 species of shrews found in the same area are also distributed over a range of almost three orders of magnitude (Brosset, 1988). The 35 species of Barro Colorado bats are also graded by size. In the three guilds of groundstory frugivores, canopy frugivores, and piscivores, the body weights differ by a factor of 1.3 to 1.8, with but one exception (Bonaccorso, 1979). As for the nine species of gleaning omnivores, they are also graded by body weight, and even more regularly by the size of their food-processing apparatus (Humphrey et al., 1983). Can we generalize these findings to all mammals in rainforest communities or to each guild within a given community? Probably not. But the parameters that can help co-occurring species to partition their resources are many. For the Barro Colorado "surface-gleaning" bats only, Humphrey et al. consider that there are as many as eight variables related to food alone.

References

Beehler, B. 1982. Ecological structuring of forest bird communities in New Guinea. In: Gressitt, J.L. (ed.), Biogeography and Ecology of New Guinea, vol. 2. Junk, The Hague, pp. 837–861

Bellier, L. 1967. Recherches écologiques dans la savane de Lamto (Côte d'Ivoire): Densités et biomasses de petits mammifères. Terre Vie 21, 319–329

Bellier, L. 1974. Application de l'analyse des données à l'écologie des rongeurs de la savane de Lamto (Còte d'Ivoire). Thèse, Université de Paris VI

Bonaccorso, F.J. 1979. Foraging and reproductive ecology in a Panamian bat community. Bull. Florida State Mus., Biol. Sci. 24, 359–408

Bourlière, F. 1983. Species richness in tropical forest vertebrates. Biol. Int. Special Issue n°6, 49–60

Bourlière, F. 1985. Primate communities: Their structure and role in tropical ecosystems. Int. J. Primatol. 6, 1–26

Bourlière, F., Minner, E. and Vuattoux, R. 1974. Les grands mammifères de la région de Lamto, Côte d'Ivoire. Mammalia 38, 433–447

Bourquin, O., Vincent, J. and Hitchins, P.M. 1971. The vertebrates of the Hluhluwe Game Reserve Corridor (state land) – Umfolozi Game Reserve. Lammergeyer 14, 5–58

Brosset, A. 1966. Recherches sur la composition qualitative et quantitative des populations de vertébrés dans la forêt primaire du Gabon. Biologia Gabonica 2, 163–177

Brosset, A. 1988. Le peuplement de mammifères insectivores des forêts du nord-est du Gabon. Rev. Ecol. (Terre Vie) 43, 23–46

Charles-Dominique, P. 1975. Nocturnality and diurnality. An ecological interpretation of these two modes of life by an analysis of the higher vertebrate fauna in tropical forest ecosystems. In: Luckett W.P. and Szalay, F.S. (eds.), Phylogeny of Primates. Plenum, New York, pp. 69–88

Crespo, J.A. 1982. Ecologia de la comunidad de Mamiferos del Parque Nacional Iguazu, Misiones. Rev. Mus. Argent. Cienc. Nat. Bernardino Rivadavia, Ecologia 3, 45–162

Davis, D.D. 1962. Mammals of the lowland rainforest of north Borneo. Bull. Nat. Mus., Singapore 31, 1–129

Dieterlen, F. 1967. Eine neue Methode für Lebendfang, Populationsstudien und Dichtebestimmungen an Kleinsäugern. Acta Tropica 24, 244–260

Dubost, G. 1987 (1988). Une analyse écologique de deux faunes de mammifères forestiers tropicaux. Mammalia 51, 415–436

Duplantier, J.M. 1983. Les Rongeurs forestiers du Nord-Est du Gabon. Thèse, Université de Montpellier

Ecotrop. 1979. Liste des vertébrés de la région de Makokou, Gabon. Laboratoire ECOTROP/CNRS, Paris. Mimeographed report

Eisenberg, J.F. and Gould, E. 1970. Appendix A. Smiths. Contrib. Zool. 27, 127

Eisenberg, J.F. and Thorington, R.W. Jr. 1973. A preliminary analysis of a neotropical mammal fauna. *Biotropica* 5, 150–161

Eisenberg, J.F., O'Connell, M.A. and August, P.V. 1979. Density, productivity, and distribution of mammals in two Venezuelan habitats. In: Eisenberg, J.F. (ed.), Vertebrate Ecology in the Northern Neotropics, Smithsonian Institution, Washington, D.C., pp. 187–207

Emmons, L.H. 1980. Ecology and resource partitioning among nine species of African forest squirrels. Ecol. Monogr. 50, 31–54

Emmons, L.H. 1984. Geographic variation in densities and diversities of non-flying mammals in Amazonia. Biotropica 16, 210–222

Emmons, L.H., Gautier-Hion, A. and Dubost, G. 1983. Community structure of the frugivorous-folivorous forest mammals of Gabon. J. Zool., London 199, 209–222

Emmons, L.H. and Gentry, A.H. 1985. Tropical forest structures and the distribution of gliding and prehensile-tailed vertebrates. Am. Nat. 121, 513–524

Erard, C. 1986. Richesse spécifique de deux peuplements d'oiseaux forestiers équatoriaux: une comparaison Gabon-Guyane. Mémoires du Muséum National d'Histoire Naturelle, Série A., Zoologie 132, 53–66

Fleming, T.H. 1973. Numbers of mammal species in North and Central American forest communities. Ecology 53, 555–563

Fleming, T.H., Hooper, E.T. and Wilson, D.E. 1972. Three Central American bat communities: Structure, reproductive cycles, and movement patterns. Ecology 53, 555–569

Foresta, H. de, Charles-Dominique, P., Erard, C. and Prevost, M.F. 1984. Zoochorie et premiers stades de la régénération naturelle après coupe en forêt guyanaise. Rev. Ecol. (Terre Vie) 39, 369–400

Gautier, J.P. and Gautier-Hion, A. 1969. Les associations polyspecifiques chez les Cercopithecidae du Gabon. Terre Vie 23, 164–201

Gautier-Hion, A., Quris, R. and Gautier, J.P. 1983. Monospecific vs. polyspecific life: A comparative study of foraging and antipredatory tactics in a community of *Cercopithecus* monkeys. Behav. Ecol. Sociobiol. 12, 325–335

Gressitt, J.L. and Nadkarni. 1978. Guide to Mt. Kaindi. Background to Montane New Guinea Ecology. Wau Ecology Institute, Handbook 5, 1–135

Hallé, F. and Oldeman, R.A.A. 1970. Essai sur l'architecture et la dynamique de croissance des arbres tropicaux. Masson, Paris

Harrison, J.L. 1959. Animal populations in rainforest. Proc. First All-India Congr. Zool. 2, 234–244

Heim de Balsac, H. 1974. Les insectivores de Lamto (Côte d'Ivoire). Mammalia 38, 637–646

Humphrey, S.R., Bonaccorso, F.J. and Zinn, T.L. 1983. Guild structure of surface-gleaning bats in Panama. Ecology 64, 284–294

Jacobsen, N.H.G. 1977. An annotated checklist of the amphibians, reptiles and mammals of the Nylsvley Nature Reserve. S. Afr. Nat. Scient. Progr., Reports 21, 1–65

Kerle, J.A. and Burgman, M.A. 1984. Some aspects of the ecology of the mammal fauna of the Jabiluka area, Northern Territory. Aust. Wildl. Res. 11, 207–222

Koford, C., Smythe, N., Hayden, J. and Bonaccorso, F.J. 1977. Checklist of Mammals of Barro Colorado Island. Smithsonian Institution, Washington, D.C. Mimeographed report

Koopman, K.F. 1979. Zoogeography of mammals from islands off the northeastern coast of New Guinea. Am. Mus. Novit. 2690, 1–17

Laval, R.K. and Fitch, H.S. 1977. Structure, movements and reproduction in three Costa Rican bat communities. Occ. Pap. Mus. Nat. Hist., Univ. Kansas 69, 1–27

Leigh, E.G. Jr., Rand, A.S. and Windsor, D.M. 1982. The Ecology of a Tropical Forest. Seasonal Rhythms and Long-Term Changes, Smithsonian Institution, Washington, D.C.

Lim Boo Liat, Muul, I. and Chai Kah Sin. 1974. Zoonotic Studies of Small Animals in the Canopy Transect at Bukit Lanjan Forest Reserve, Selangor, Malaysia. Paper presented at the IBP Meeting, Kuala Lumpur, 12–14 August, 1974

McKenzie, N.L. and Rolfe, J.K. 1986. Structure of bat guilds in the Kimberly mangroves, Australia. J. Anim. Ecol. 55, 401–420

Mares, M.A., Willig, M.R., Streilin, K.E. and Lacher, T.E. Jr. 1981. The mammals of Northeastern Brazil: A preliminary assessment. Ann. Carnegie Museum 50, 81–137

Medway, Lord and Wells, D.R. 1971. Diversity and density of birds and mammals at Kuala Lompat, Pahang. Malay. Nat. J. 24, 238–247

Menzies, J.I. and Dennis, E. 1979. Handbook of New Guinea rodents. Wau Ecology Institute, Handbook 6, 1–68

National Research Council. 1981. Techniques for the Study of Primate Population Ecology, National Academy Press, Washington, D.C.

Nowak, R.M. and Paradiso, J.L. 1983. Walker's Mammals of the World, 4th ed. John Hopkins University Press, Baltimore

Patton, J.L., Berlin, B. and Berlin, E.A. 1982 Aboriginal perspectives of a mammalian community in Amazonian Peru: Knowledge and utilization patterns among the Aguarana Jivaro. In: Mares, M.A. and Genoways, H.H. (eds.), Mammalian Biology in South America, Pymatuning Laboratory of Ecology, Pittsburgh, pp. 111–120

Payne, J.B. 1979. Synecology of Malayan Tree Squirrels, With Special Reference to the Genus Ratufa. Ph.D. thesis, University of Cambridge

Pienaar, U. de V. 1963. The large mammals of the Kruger National Park. A systematic list and zoogeography. Koedoe 7, 1–25

Pienaar, U. de V. 1964. The small mammals of the Kruger National Park. A systematic list and zoogeography. Koedoe 7, 1–25

Pine, R.H., Bishop, I.R. and Jackson, R.L. 1970. Preliminary list of mammals of the Xavantina/Cachimbo expedition (central Brazil). Trans. R. Soc. Trop. Med. Hyg. 64, 668–670

Poulet, A.R. 1972. Recherches écologiques sur une savane sahélienne du Ferlo septentrional, Sénégal. Les mammifères. Terre Vie 26, 440–472

Rautenbach, I.L., Fenton, M.B. and Braack, L.E.O. 1985. First records of five species of insectivorous bats from the Kruger National Park. Koedoe 28, 73–80

Richard, A.F. 1985. Primates in Nature. Freeman & Co., New York

Schaller, G.B. 1983. Mammals and their biomass on a Brazilian ranch. Arquiv. Zool. 31, 1–36

Sourd, C. and Gautier-Hion, A. 1986. Fruit selection by a forest Guenon. J. Anim. Ecol. 55, 235–244

Struhsaker, T.T. 1981. Polyspecific associations among tropical rainforest primates. Zeits. Tierpsychol. 57, 268–304

Sussman, R.W. and Raven, P.H. 1978. Pollination by lemurs and marsupials: An archaic coevolutionary system. Science 200, 731–736

Terborgh, J. 1983. Five New World Primates: A Study in Comparative Ecology, Princeton Univ. Press, Princeton, NJ

Terborgh, J., Fitzpatrick, J.W. and Emmons, L. 1984. Annotated checklist of bird and mammal species of Cocha Cashu Biological Station, Manu National Park. Fieldiana, Zoology 21, 1–29

Thomas, D.W. 1982. The Ecology of an African Savanna Fruit Bat Community: Resource Partitioning and Seed Dispersal. Ph.D. thesis, University of Aberdeen

Whitten, J.E.J. 1981. Ecological separation of three diurnal squirrels in tropical rainforest on Siberut Island, Indonesia. J. Zool., London 193, 405–420

Wilson, D.E. 1983. Checklist of mammals. In: Janzen, D.H. (ed.), Costa Rican Natural History, University of Chicago Press, Chicago, pp. 443–447

Ziegler, A.C. 1982. An ecological checklist of New Guinea recent mammals. In: Gressitt, J.L. (ed.), Biogeography and Ecology of New Guinea, Junk, The Hague, pp. 863–894

7. Species Diversity in Tropical Vertebrates: An Ecosystem Perspective

François Bourlière and Mireille L. Harmelin-Vivien

Most of the hard evidence and current views pertaining to the causes of the high vertebrate species diversity in tropical coral reefs and rainforests have been reviewed in the preceding pages, and their implications at the community level discussed.

However, the understanding of the ways in which species-rich communities are structured and maintain themselves does not forcibly tell us why such assemblages are the rule in the humid tropics but not elsewhere; some further considerations on this subject might therefore be timely.

In these concluding remarks we will first summarize and compare the attributes of the two environments in which the richest vertebrate species assemblages are found, at sea and on land. To bring into focus the differences as well as the similarities between coral reef and rain forest, and between taxonomic groups, will help to better evaluate the respective roles of some of the factors maintaining species richness at the ecosystem level. In doing so, we will have to remember that in both environments vertebrates are not the only (nor the major!) consumers of plant and animal resources. The many thousands of invertebrates must always be kept in mind, although their trophic impact has seldom been estimated accurately.

Next, we will try to demonstrate that high species diversities in animals can neither be explained within the limits of a single taxonomic category, nor even at the level of a single trophic group. As already foreseen by some authors, the increase in animal species diversity* within an ecosystem is a multistage process, which has

---

*Following Whittaker (1977), species richness and species diversity are considered as synonymous in the present chapter.

its origin at the primary production level. It is here that one must seek out the true causal factor(s).

Rainforest and Coral Reef Compared and Contrasted

Before proceeding any further in our comparison of species diversity in tropical rainforests and coral reefs, it is imperative to emphasize once more the importance of the matter of scale in such studies.

Rainforests and coral reefs are now considered generally as open, nonequilibrium systems (Connell, 1978; Price, 1984) rather than steady-state systems, as they were in the past. All tropical rainforests are indeed patchworks of different forest facies, and of various stages of regeneration of forest gaps following tree falls or similar stochastic accidents. In the same way, coral reefs are mosaics of coral patches of different ages and architectures, as a consequence of frequent storms and occasional hurricanes. It is therefore necessary, when one discusses species diversity in these two systems, to distinguish between the overall species richness of a forest or a whole reef tract, and that of much smaller areas, such as a 1-ha forest plot or an isolated patch reef. In other terms, one should never equate the structure of a landscape with that of a small-size study area, whether it be above or below water level.

This distinction between broad-scale and small-scale studies is even more important in the marine environment than on land. Whereas many (if not most) tropical rainforest vertebrate species are sedentary, or locally nomadic at the most, only a few coral reef fishes spend their early months of life in the same habitat as their parents. Eggs and young fry drift away to be carried out by oceanic currents, and only a large reef tract can expect to contain at any time most of the representatives of the available species pool. On the contrary, the recruitment of young fish on smaller sites being largely stochastic, only a part of this species pool will be able to settle on small reef patches, and the fish communities of these patches will therefore vary greatly in space and time (Sale, 1984; and Chapter 1 of this volume).

In the following pages, we will be concerned mainly with species diversity at its broadest scale, i.e., that of the ecosystem, as it is the level at which the greatest temporal constancy is achieved—so long as the environment does not change drastically.

Table 7.1 summarizes the major physical and biological attributes of the two ecosystems concerned and makes comparisons easier. The similarities and differences of the two systems stand out immediately, the former being much more numerous than the latter.

In both rainforest and coral reef, habitat diversity provides a variety of space resources to their animal dwellers. There is no need to elaborate any further here on the extreme structural heterogeneity of the vegetation in tropical rainforests, or on the large number of microhabitats made available for all size-class of vertebrates and invertebrates (most of them of small size, contrary to a widely held view). This is made possible not only by the layering of the arboreal vegetation, but also by the varied growth habits and "architectural models" of tropical forest trees, and by the abundance of lianas and epiphytes, which have no real counterpart in temperate forests, whether broadleaved or coniferous.

Table 7.1. Ecosystem Attributes of Particular Importance to Vertebrates in Rainforests and Coral Reefs

	Rainforests	Coral Reefs
Physical Attributes		
Physical environment	Air	Sea water
Climate:		
Temperature	Stable, moderate (average 25–28 °C)	Stable, moderate (average 20–25 °C)
Rainfall	High (above 1500 mm)	Highly variable
Physical structure:		
Substratal relief	Lowland vs. mountain forest	Fringing vs. barrier reef and lagoon
Substratum diversity	Soil diversity	Percent of coral cover and reef sediments
Surface area	Good predictor of species numbers present	Good predictor of species numbers present
Physiognomy	Large forest blocks to gallery forest	Large coral reefs to coral patches
Indicators of stochastic "catastrophic" events	Percent of recent gaps	Percent dead coral cover
Biological Attributes		
Species richness of:		
Plant producers	High	Moderate
Animal consumers	Very high	Very high
Equitability of plant & animal communities	Low	Low
Structural diversity	A complex 3-dimensional living habitat	A complex 3-dimensional living habitat
Diversity of food resources	High	High
Phenology	Reduced seasonality	Reduced seasonality
Productivity		
Primary	High and sustained	High and sustained
Secondary	High and sustained	High and sustained
Biotic interactions		
Competition	Supposed to be strong	Supposed to be strong
Mutualistic interactions	Numerous	Numerous
Predation pressure	Likely high	Likely high
Parasite & pathogen load	Likely high	Likely high
Habitat selection by vertebrates	Mostly active	Initially passive

The same is true for coral reefs where calcareous algae and corals build up a complex tridimensional reef frame as a result of their variety of life forms and ecomorphs. This living framework is made even more intricate by the overgrowth of many algae and sessile invertebrates, and the activities of a wealth of boring organisms, which again increase the number of microhabitats available. But the actual living space to be considered to account for reef fish diversity also includes the whole water mass around the reef itself. Numerous fishes of various sizes dwell throughout the water column, from just above the reef up to the water surface. Some of the larger fish species spend most of their adult life over the reef itself, even if they may sometimes rest on it, or on the sandy bottom nearby. It is worth noticing here that there are many more large predators feeding over a reef tract than over a rainforest, where a few large birds of prey only hunt above the canopy, whereas small insectivorous bats, swifts, and swallows exploit the "aerial plankton."

The diversity of food resources—another major attribute common to tropical rainforests and coral reefs—appears to be an even more important contributory factor of their species diversity than their structural heterogeneity, at least for animals.

This is very obvious in the lowland evergreen rainforests of the tropics where the number of plant species per unit area is extremely high as compared to more northern or southern latitudes. For instance, the number of plant species reaches 365 per 0.1 ha in the rain forest of Rio Palenque, Ecuador (Gentry and Dodson, 1987). As each plant species provides a "complex of food niches," to use Harper's words (1977), there will inevitably be more species of animal primary consumers than of vascular plant species. A striking example of this multiplying effect is provided by Erwin's recent studies of the canopy beetles living on a single species of forest tree of Panama, *Luchea seemanii* (summarized by May, 1986). By using insecticidal "fog," Erwin collected no less than 682 different species of herbivorous beetles on only 19 individual trees; 20% of them were estimated to be host-specific. In turn this profusion of insect prey attracted numerous predators, out of which there were 296 species of beetles. Moreover, 96 species of scavenging beetles and 69 fungus eaters were found in the same canopies, not to mention all the other orders of insects! In the rain forest of eastern Peru, 43 species of ants belonging to 26 genera were found on a single tree, i.e. approximately as many as the entire ant fauna of all habitats in the British Isles (Wilson, 1987). Nothing so spectacular, of course, occurs to vertebrates whose species diversity remains always much lower in tropical rainforests than that of the insects. Nevertheless, the extreme variety of size, shape, color, taste, and smell of fruits (see, for instance, Roosmalen, 1985, for the Guianas) allows the sustenance of a large number of avian and mammalian fruit consumers. Similarly, the thousands of species of insects and spiders that dwell in all corners of the forest are looked after by many insectivorous vertebrates: tree frogs, lizards, birds, bats, and even some scansorial mammals, most of them well adapted morphologically and/or behaviorally to forage in all the existing microhabitats of their tridimensional living environment. Terborgh (1985), among others, notes that the differentiation of the ecological roles of insectivorous birds in western Amazonia is based largely on the differences of search and capture behaviors which, in turn, relate to the parts of the vegetation to which the hunting insectivore directs its attention.

The variety of food resources in a coral reef is equally striking. Although there are far fewer species of plants in the sea than on land, as the number of terrestrial species of angiosperms is about 10 times larger than that of algae, there are more kinds of marine plants in tropical than in temperate marine communities. For example, more than 50 species of algae were identified on a few square centimeters of reef algal turf in the West Indies (Wanders, 1976). On the other hand, the patterns of primary production and the plant-herbivore interactions are quite different in a coral reef and in a tropical rainforest. The reef primary producers include phytoplankton, benthic macro- and microscopic algae, and symbiotic algae. Yet phytoplankton production is low in tropical waters, and not directly used by reef fishes, whereas benthic primary production on reefs is one of the highest recorded (Lewis, 1977). As for the primary production of symbiotic algae associated with corals and other sessile invertebrates, it is even higher than that of the benthic algae (Larkum, 1983). On the consumers' side, most of the reef herbivores are generalist feeders, grazing on a variety of benthic algae, although the percentage of specialist algal feeders might well increase as studies of fish dietary preferences become more numerous and elaborate (Ogden, 1985). However, it appears that, more than their algal species richness, it is their common occurrence, their high productivity, and their palatability which make the algal turfs so attractive for so many reef herbivores.

The number of invertebrate species on a coral reef far exceeds that of the plants, as in a rainforest. For example, there are about 350 species of benthic algae so far recorded on French Polynesian reefs, vs. more than 2000 species of invertebrates (Richard, 1985); probably as many remain to be discovered. Polychaete worms, crustaceans, and molluscs are the more species-rich groups. For instance, 103 worm species were found on a single coral head in the Great Barrier Reef (Grassle, 1973). Most of these invertebrates are of very small size and thrive into all reef microhabitats, often associated with other living organisms. However, it must be pointed out here that no category of reef invertebrates does ever match the high species diversity of some insect groups in the humid tropics. The number of species of Coleoptera and Diptera alone might far exceed the million (Erwin, in May, 1986). The reef invertebrates are the staple food of a number of carnivorous species and, among fishes, the diversity of carnivores always exceeds that of herbivores. These carnivorous fishes exhibit a great variety of morphological and behavioral adaptations to their favorite preys. Emery (1973), for example, showed how the feeding habits of 14 species of pomacentrid fishes influenced their morphological characteristics in relation to the vertical distribution of the fishes in the water column.

The year-long availability of food and space resources in both rainforest and coral reef is as important as their variety to maintain high animal species diversities. Although seasonality in flowering and fruiting appears to be the rule in all rainforests—even the few reputed to grow under "aseasonal" conditions—the major food categories remain available throughout the year. They may vary in nature and amount, but are always there, or not very far away. Furthermore, the periods of relative scarcity are predictable, and always of rather short duration; they are never comparable in intensity and duration to what is the rule in dry woodlands or savannas, not to mention the harsh long winter of the higher latitudes. Moreover, in tropical rainforests there are always some "keystone plant resources" (Leighton and Leighton, 1983; Terborgh, 1986), such as figs, palm nuts, nectar, or gums, that allow vertebrate frugivores, for instance, to cover their nutritional requirements through the period of scarcity. This implies that these frugivores may have to temporarily pursue alternative life-styles in order to survive, and also that their populations may in this way be kept down to levels compatible with the food resources available during these "lean periods." But these animals do not need to move away and can successfully face these temporary adverse conditions. We still do not know whether there are "keystone food resources" in coral reefs. But a peculiar and important food source, of high caloric value, is made up by the billions of eggs periodically released in the water by reef invertebrates and fishes.

Unfortunately, this has not yet been quantified. In any case, most carnivorous and omnivorous organisms gorge themselves on coral eggs during the mass spawning of scleractinians on the Great Barrier Reef (Babcock et al., 1986). Concentrations of predators prowling around spawning grounds are in fact reported from all aquatic environments (Lowe-McConnell, 1987).

The moderately high environmental temperature, which prevails all year round, and is incidentally the climatic parameter that changes the least from month to month, also has other advantages. Combined with a high atmospheric humidity, it allows ectothermic vertebrates and many invertebrates to remain active continuously. The benefit is even greater for endothermic birds and mammals. Not having to use a lot of calories to keep their body temperature at constant high level, they save a good deal of energy and can content themselves with low-calorie diets, such as fresh leaves or mushrooms. This might also help to explain why so many herbivorous fish species are observed in coral reefs, as well as herbivorous and detritivorous species in tropical freshwater fish communities (Lowe-McConnell, 1987), whereas these trophic categories are rare in temperate waters.

The increased primary productivity of the humid tropics has in itself been considered as an important contributory factor in the higher species diversity of sympatric animal communities. It has been supposed for a long time that tropical rainforests are indeed the most productive type of natural vegetation on land, and the coral reefs have often been considered as "oases of flourishing life in the watery desert of tropical oceans." But such statements need to be tempered. It is true that high solar energy inputs and tolerably high temperatures in a humid atmosphere do stimulate photosynthesis during the day, but as this temperature remains also high during the night in equatorial latitudes, a good deal of what has been photosynthesized during the day is then lost through respiration. As a consequence, the net primary production (i.e., the plant tissues that can be used for food by animals) always remains lower than its gross production. The same happens in a coral reef. This explains why the amount of dry matter produced per m^2 and per year in a tropical rainforest, for example, remains less than twice that of a temperate evergreen forest, whereas its plant species diversity may be 10–100 times greater. It must be emphasized that if a higher net primary production does favor higher animal species richness in the humid tropics, this is not so much because rain forests and coral reefs produce far more of a same kind of food than equivalent ecosystems in higher latitudes, but because they produce a much wider range of food sources in larger amounts, and on a year-long basis (Begon et al., 1986).

The role and importance of biotic interactions such as interspecific competition, predation pressure, or mutualism as other contributory factors to the maintenance of high animal species diversities in the humid tropics has been often—and sometimes hotly—discussed during recent years. That such interactions should be more numerous in tropical communities than at higher latitudes is indeed likely, if only because there are so many more species permanently sharing a same "volume of habitat" in a rainforest or a coral reef than in other ecosystems. But to what extent some of these interactions can be considered as causes, rather than results, of high species diversities often remains unsettled.

In any case, the low equitability of plant and animal communities in tropical rain-forests and coral reefs cannot but help to reduce competition among species-rich assemblages and contribute to their maintenance. Among vertebrates and most inver-tebrates, species are unevenly represented in humid tropics communities; a few are relatively abundant, whereas most of the others are scarce. Whether or not a heavy pre-dation pressure contributes to this state of affairs is still a matter of debate.

Some ecosystem attributes, however, are strikingly different in tropical rainforests and coral reefs. The most obvious difference, of course, relates to the nature of the physical environment of plant and animal organisms: a hot and humid atmosphere in the forest, a hot but well-oxygenated seawater in the reef. There may be some excep-tions to this sharp dichotomy, such as the seasonal flooding of some riverine forests and floodplains, and the daily exposure at low tide of parts of the reef flat, but they remain relatively minor.

The second difference is more apparent than real. The structural diversity of the rain forest results from the heterogeneity of the vegetation and the highly diverse architec-ture of the trees, whereas the complex structure of the coral reef is due to the variety of growth forms of hard corals and other sessile invertebrates. But this difference is misleading; as previously mentioned, the algal component of scleractinians is metabol-ically so active that it provides most of the energy requirements of these invertebrates, even though the bulk of the reef is made of dead coral skeletons, as indeed the bulk of a tropical rainforest is mostly made up of the physiologically inactive heart wood of the tree trunks!

A more important difference between rainforests and coral reefs relates to the trophic structure of the vertebrate communities in the two systems. Whereas most of the rain-forest mammals and birds are primary consumers, and amphibians and reptiles secon-dary consumers, no more than 10–15% of the coral reef fishes are "herbivores" when adults (Harmelin-Vivien, Chapter 2 of this volume).

Quite likely, the most important difference between the rain forest and coral reef vertebrate communities lies in the different modalities of recruitment of their re-spective populations. Whereas most of the forest amphibians, reptiles, birds, and mammals are sedentary, their young being born and raised locally, most of the coral reef fishes, as already mentioned, have lengthy pelagic larval phases followed by a sedentary reef-associated adult phase that may last for years. This means that, at a small scale, local assemblages are formed through the successive settlement of new juvenile individuals of various species (Sale, 1985, and Chapter 1 of this volume). But the species composition of fish communities, even in a small reef tract, can-not be considered as entirely stochastic. No fish species can settle and spend its adult life anywhere within a reef. Many are restricted as adults to very specific microhabitats, to which they may remain associated for months if not years. For instance, Eckert (1985) showed experimentally that, already at settlement, many species of coral reef fish have definite preferences for certain features of habitat. It nevertheless remains true that habitat selection in terrestrial rainforest vertebrates is always an active process, while it is mostly passive among coral reef fishes, at least for the early larval stages of many species. However important these differ-ences in patterns of recruitment might be, they apparently have little influence on the number of species able to co-occur and breed successfully in a same tropical environment.

Why Species-Rich Communities Cannot Exist in Highly Seasonal Climates

If the great variety of animal species of the humid tropics is made possible by the wider range of food and habitat resources these ecosystems provide throughout the year, and in a greater amount than elsewhere, then the species diversity must decrease any time a harsh seasonal climatic episode sharply reduces, or completely interrupts, organic production for part of the year. This is what happens as we move away from the humid tropics. In tropical savannas, both primary and secondary production—as well as the species diversity of most plant and animal groups—generally decline in proportion with the severity and duration of the dry season. In the so-called temperate climates, the length of the winter period and the depth of the snow cover play a similar role. At even higher latitudes, the period of organic production may even be restricted to a few weeks, and the resident vertebrate communities are limited to very few species.

It is not only the production of food that is decreased during these "lean" periods of the year; the access to what little food remains available on or below ground (seeds, resistance stages of invertebrates, ground vegetation, etc.) is hampered, either because many invertebrates took refuge in the soil or because a deep cover of snow prevents most vertebrate consumers from locating them. No wonder, therefore, that in each kind of highly seasonal environment, whether it be tropical or "temperate," the number of resident animal species is always lower than in the humid tropics, and that their populations remain below their maximum possible levels because the amount of food, and the availability of cover, are limited during the "lean" season. Consequently, part of the local organic production can even be left unused by resident vertebrates. Such seasonal energy surpluses are then used by migratory or nomadic species (Bourlière, 1961; Morel and Bourlière, 1962).

The opportunities open to resident vertebrate species to successfully face periods of scarcity in highly seasonal climates are indeed limited. When food and cover are drastically reduced, there are only two possibilities left (other than to die out!):

One strategy is to move for a time, at the lowest possible energy cost, to other kinds of environments where some extra food is still available. This is what migrants, both short- and long-distance, do.

The other strategy is to enter some sort of "resistance stage" that enable individuals to stay alive without consuming any food at all (except the energy stored in their own bodies). Many invertebrates have their life cycles adjusted to the local climatic conditions in such a way that they spend the dry season, or the winter, in diapause at the egg or the pupal stages—as seeds of plants enter dormancy. Larger animals, such as vertebrates, have in similar conditions to rely on some kind of lethargic stage (estivation, hibernation, etc.) during which they live on the energy stored during the "fat" period of the year. This is what happens to some rodents and bats and, to a lesser extent, to amphibians, reptiles, and even freshwater fishes; thus some Siberian species can "dig themselves into the mud and fall in a state of torpor" during the long winters; some coastal marine fishes may in the same circumstances move to deep water to overwinter (Nikolsky, 1963).

It has sometimes been asked why fewer species became adapted to the higher latitude environments than to those of lower latitudes (Blondel, 1986, p. 20). The answer is that one does not turn overnight into a long-distance migrant or a hibernator! In both strategies, very special adaptations are required, locomotor and metabolic, that have taken a very long time to develop. Therefore, all the taxa morphologically and physiologically unable to do so were obligatorily prevented from entering temperate and boreal latitudes, and settling there.

The vertebrate faunas of nontropical latitudes are therefore far from being a mere random subset of the species pool of the humid tropics—and the same is true for invertebrates and plants. Only the taxa able to withstand, one way or another, prolonged periods of seasonal harshness can live permanently at higher latitudes. The endothermic phyla, birds and mammals, were therefore at a definite advantage when compared to amphibians and reptiles, which cannot stand prolonged low temperatures, nor low atmospheric humidity for amphibians. This is why the latitudinal gradient in species richness is much steeper for lower vertebrates than for birds and mammals. Furthermore, all groups depending on resource categories not available in sufficient amounts throughout the year in temperate and boreal latitudes were forcibly excluded from their resident contingent of vertebrates. This explains the lack of arboreal folivores, and the extreme scarcity of frugivores, among sedentary vertebrates outside the humid tropics. On the contrary, groups able to make use of some sort of stored energy, like seeds or dry grass, are at an advantage.

The resident species of the humid tropics are so closely adapted to their unique environments that they appear quite unable to settle elsewhere, even in the nearby seasonal savannas. A case in point is that of the intertropical migrants among birds. None of them is indeed a true rainforest bird. All are savanna species taking advantage of the seasonal flushes of insect food taking place in other savanna areas, at the time when food is becoming scarce in their breeding grounds. This does not imply that all rainforest birds are sedentary; as mentioned by Erard (Chapter 4 of this volume), a number of large forest frugivores do wander extensively within the forest to feed on their patchily distributed food. Nevertheless none of them ever wanders for long outside the forest edges.

Contributory and Causal Factors of Tropical Species Richness

Any existing biotic diversity is the end result of a balance between the positive effect of speciation and the negative effect of extinction (Baker, 1970). If we were able to establish that the environmental conditions prevailing in the humid tropics are not only able to enhance the survival of the many already existing taxa but also increase the rate of speciation, then the relative role of the various determinants of species diversity would be easier to evaluate. Besides the factors that unquestionably contribute to the maintenance of a high species diversity of vertebrates in the humid tropics, we must therefore look for those environmental parameters that could be capable of increasing the rate of speciation in this environment.

Another general remark must also be made before proceeding further. When species diversity is considered at the ecosystem level, and not merely at the level of one or another taxonomic category, one general pattern emerges. Broadly speaking, animal species diversity at a given trophic level is essentially made possible by the variety and the pattern of availability of the food and space resources provided by the trophic level just below. The number of invertebrate and vertebrate carnivores, insectivores, parasitoids, and parasites clearly depends in any ecosystem on the variety and availability of the animal prey. In turn, these prey animals are for their most part supported by a large number of plant species, small and large, live or dead, affording food and/or cover to a large number of consumers. Diversity in the vegetable kingdom can only beget diversity in the animal kingdom. This is made clear by experimental or accidental manipulations of natural communities. When the plant species richness of a rain forest is decreased by selective logging or tree poisoning, the species richness of the associated fauna quickly decreases (Leighton and Leighton, 1983; Holloway, 1987; Driscoll and Kikkawa, Chapter 5 of this volume). Not only is this due to a qualitative decrease of the food resources range, but the structural diversity of the forest is also altered together with its spatial diversity on many scales.

The situation differs only in appearance in a coral reef. Its structural and dietary diversity results from a mosaic of algal turfs, patches, and ridges of coralline algae, and an abundance of scleractinian corals of varied growth forms and architectural models, and their associated invertebrates. However, the usual metabolic predominance of plants is maintained within the coral reef biotic community. Hard corals as well as many gorgonians, sponges, and even molluscs harbor in their tissues symbiotic algae that can provide as much as 90–108% of their total energy requirements (Davies, 1977; Hatcher, 1988).

In tropical freshwaters, the diversity of plant food sources is not at once obvious either. However, many species-rich fish faunas of the neo- and paleotropics (more than 700 species of fish occur within 20 km of Manaos, for example) do subsist on plant material as much as on animal food. This is made possible by an apparent lack of specialization in foods eaten by many of the fishes. Many species, when adult as well as young, feed on freshwater plankton, aufwuchs, vegetable debris, detritus trapped in roots of "floating meadows," and submerged hydrophytes. In the headwater forest streams of the Amazon and Zaire basins, where there is no phytoplankton, fish subsist almost exclusively on allochtonous material falling from the forest canopy (flowers, pollen, fruits, seeds, leaves, insects and spiders, etc.). In the lower reaches of the larger rivers, other herbivorous species also eat bottom detritus and detrital mud (Lowe-McConnell, 1975 and 1987; Goulding, 1980; Brosset, 1982; Araujo-Lima et al., 1986). In this case, living space and spawning sites are more limiting than food.

As for the amphibians and reptiles which are for their most part carnivorous, their species richness in the humid tropics can be better explained by the abundance of their animal preys than by their diversity (Bellairs, 1969; Duellman and Trueb, 1986; Duellman, Chapter 3 of this volume). Although quantitative studies of diets and feeding habits in these two vertebrate phyla remain unfortunately scarce in the tropics, it however appears that most amphibians and reptiles are also rather indiscriminate feeders, whose diet is more influenced by prey availability and prey/predator size ratios than by a deliberate selection of specific food items by the predators. However, there are exceptions, particularly among snakes (Toft, 1985; Duellman, Chapter 3 of this volume).

It is therefore quite likely that the ultimate determinants of the high species richness of all animal consumers in the humid tropics are to be traced in the high and sustained primary productivity of the ecosystems concerned, and in their floral diversity, both on land and in water.

However, the very origin of the floral diversity of the humid tropics still remains to be explained. Whittaker (1977) once remarked that "diversity patterns in plants seem at cross purposes with those in birds; scarcely a relationship applies in the same way to both." Indeed, contrary to what seems the case for animal consumers, plant species diversity cannot be accounted for by the multiplicity of food resources available to them. Whatever the latitude, all plants use the same three basic resources—light, water, and nutrients—differing only in the proportions they need of each. Unlike animals, there is therefore little opportunity for different species to specialize on different resources (Begon et al., 1986).

In the equatorial belt of our Earth, light is supplied at a high rate, and evenly throughout the year. Furthermore, both in the rainforest and in the coral reef, light is reflected and diffused down a long column of vegetation or of water. Hence, there is not only a high rate of supply, but also gradients of light intensities and a wide range of frequency spectra. This leads to equally wide ranges of light regimes and to increased opportunities for specialization, and thus to an increased species diversity (Begon et al., 1986).

Water is seldom if ever a limiting factor in the humid tropics either, but plant species richness in such latitudes is nonetheless strongly influenced by the amount and regime of the rains. The study of neotropical plant species diversity carried out by Gentry (1982) clearly shows that precipitation is strongly correlated with plant community diversity and organization. In general, wetter sites have more diverse plant communities. This trend holds for all synusia (habit classes) taken separately, but the sensitivity of the different synusia to changes in precipitation differs. Epiphyte diversity increases most rapidly with precipitation, diversity of large trees least rapidly, lianas and herbs at intermediate rates. For any given neotropical plant community, diversity, at least of 1000-m^2 samples, can be accurately predicted from rainfall data.

The nutrient situation is more complex (Vitousek and Sanford, 1986). Many tropical rainforest grow on relatively poor soils, and this situation apparently does not hamper their plant species richness. Indeed, the data reviewed by Tilman (1982) for several Malayan rain forests suggest that richness may be highest at intermediate levels of phosphorus and potassium concentration in the soils. Although more studies are obviously needed, such "humped relationships" (Begon et al., 1986) might also explain why high plant species diversities are also often found on poor soils outside rainforests, as it is for example the case for the cerrado vegetation in Brazil, where species richness of up to 230–250 species per 0.1 ha have even been reported (Silberbauer-Gottsberger and Eiten, 1983). As a matter of fact, experimental resource enrichment often leads to a decrease in species richness in plants (the "enrichment paradox"). This is particularly striking for the phytoplankton: in coastal marine regions, as in lakes and estuaries, eutrophication results in a decrease in the diversity of phytoplankton and an increase in primary productivity (Tilman, 1982).

A fourth important environmental parameter of the humid tropics remains to be considered, i.e., the constant and moderately high air and water temperature that, within physiological limits, is well known by all biologists for its multifarious influence on most life processes. This effect is all the more important in organisms, plant or animal, that are ectothermic.

Could not, then, the tolerably high and constant temperature of the humid tropics be the single extrinsic factor that "sets the ball rolling," to which some contemporary ecologists refer in the discussion on species richness (Begon et al., 1986; Lavelle, 1986)? Could not this greenhouse atmosphere, coupled with a high rate of solar energy input operating over millions of years, have also influenced the rate of organic evolution?

A number of experimental results now make such an hypothesis plausible. We've long known that, within the temperature range that permits active life, an increasing temperature speeds up the rate of enzymatic reactions, including at the DNA level. Even more important, there is now a distinct possibility that temperature could exert an effect on the very structure of the genome itself (Bernardi and Bernardi, 1987). On the other hand, it has also been long known that a high environmental temperature can also increase the rate of mutation (Serra, 1968; Lindgren, 1972a, 1972b; Auerbach, 1976). This is well established for bacteria and ascomycetes. A positive temperature–mutation rate relationship is also likely in vascular plants but more difficult to interpret in *Drosophila* and other invertebrates.

If temperature does in fact affect the structure and the stability of the genome (at its DNA, chromatin, and chromosome levels), the environmental conditions peculiar to the humid equatorial zones can reasonably be held responsible for increasing the rate of speciation in microorganisms, plants, and ectothermic invertebrates and vertebrates. At the very end of this "cascade reaction," endothermic birds and mammals would benefit indirectly, through the variety of resources provided by plants and ectothermic animals, and through the spatial heterogeneity on many scales of rainforests and coral reefs.

Conclusions

There is obviously no simple explanation for the high animal species richness in the humid tropics. No single factor can be held responsible for enhancing and maintaining a high species diversity in this environment. However, some suggestive patterns begin to emerge from the evidence available, pointing toward a hierarchy of causal factors.

(1) Within the coral reef and the rainforest, the maintenance of a high animal species diversity mostly depends on the steady availability of a wide and diverse array of food resources. These resources are provided, the year long and in a predictable way, both by the plant and animal components of the ecosystems. Such a large spectrum of resources allows an extreme variety of life forms and life-styles to be adopted by a large number of species—many of them becoming more or less narrowly specialized. In the

last analysis, the prime cause of the high species richness of all animal consumers, whether they be vegetarians, carnivores, or mixed feeders, appear to rest, directly or indirectly, on the more or less sustained primary productivity of both coral reefs and rainforests, and the variety of food resources produced.

(2) This fairly even level of primary production of the coral reefs and rain forests, and the consequent lack of severe and unpredictable seasonal shortages of food, is made possible in the humid tropics by peculiar climatic conditions: a constant but moderately high temperature whose daily and seasonal variations are always small and predictable, a steady but not excessive supply of solar radiation, and a high atmospheric humidity.

As soon as the climate becomes more seasonal, from tropical woodland and savannas toward the temperature latitudes, organic production starts fluctuating more and more according to the seasons, sometimes in an unpredictable way. The resident organisms, plants and animals, have then to develop very special adaptations to cope with the harsh conditions of the long dry seasons or cold winters. Their species diversity is simultaneously reduced.

(3) Both coral reefs and rainforests are tridimensional living structures made up of modular organisms (trees and sessile invertebrates), whose architectures provide a variety of habitat resources. This habitat heterogeneity enables a large number of taxonomically or ecologically related species to co-occur in a same environment. Habitat heterogeneity is again increased by the dynamic nature of the coral reefs and rainforests. Both are nonequilibrium systems made of patches of different ages representing different successional stages, which coexist in different proportions in the same seascape or landscape. Furthermore, large modular organisms (particularly large trees) can behave as actual colonies of organisms, significant phenotypic variation sometimes occurring between modules – some of it genetic in origin (Gill, 1986). Such a mosaicism cannot but contribute to increase the heterogeneity of the rainforest (and perhaps of the reef!).

(4) We still do not know whether or not the climatic conditions so favorable to the maintenance of a high species diversity in the humid tropics also play a role in increasing the rate of speciation of the plants and animals living in this environment. However, constant and moderately high temperatures (in the range of those prevailing in the ecosystems concerned) are known to accelerate the rate of living of many organisms and to influence the stability (if not the structure itself) of their genomes. It is therefore not unrealistic to think that such thermal conditions were able, over the course of time, to increase the rate of speciation of many organisms, and perhaps still continue to do so.

(5) Not enough attention has been given until now to the exceptions to the usual increase of species diversity toward the equator and to their possible causes.

First of all, it must be realized that some resources can be scarcer in the humid tropics than in the dry tropics or the temperate latitudes. In this case the organisms specialized in their utilization are less numerous in the humid tropics than elsewhere. Such is the case, for instance, for grass in rainforests; consequently, the grazing ungulates so species-rich in some savannas give way to a few browsing or frugivorous ungulates feeding on leaves or fallen fruits. Earthworms also can become very rare in some rainforests, when the leaf litter is too rapidly destroyed by fungi and/or termites; then, their communities are far less diverse than in some tropical savannas or even temperate habitats (Lavelle, 1986).

Moreover, latitudinal gradients of species diversity often differ among taxonomic categories, some of them being definitely more sensitive to some environmental parameters than to others. The paramount importance of temperature for reptiles, and of air moisture for amphibians, has been emphasized by Duellman (Chapter 3 of this volume). As far as diet is concerned, it also appears that the geologically older groups (fishes, amphibians, lizards) appear to be less concerned for the nature of their dietary items than for their size and amount. On the contrary, the quality of the food is as important, or even more important, as its quantity for the more "modern" groups (snakes, birds, mammals). There are consequently more generalist feeders and fewer specialists among the former than among the latter. The role of phylogenetic constraints here becomes obvious.

References

Araujo-Lima, C.A.R.M., Forsberg, B.R., Victoria, R., and Martinelli, R. 1986. Energy sources for detritivorous fishes in the Amazon. Science 234, 1256–1258

Auerbach, C. 1976. Mutation Research, Chapman Hall, London

Babcock, R.C., Bull, G.D., Harrison, P.L., Heyward, A.J., Oliver, J.K., Wallace, C.C., and Willis, B.L. 1986. Synchronous spawning of 105 Scleractinian coral species on the Great Barrier Reef. Mar. Biol. 90, 379–394

Baker, H.G. 1970. Evolution in the Tropics. Biotropica 2, 101–111

Begon, M., Harper, J.L., and Townsend, C.R. 1986. Ecology. Individuals, Populations, and Communities, Blackwell, Oxford, UK

Bellairs, A. 1969. The Life of Reptiles, Weidenfeld and Nicholson, London

Bernardi, G., and Bernardi, G. 1987. Compositional constraints and genome evolution. J. Molec. Evol. 24, 1–11

Blondel, J. 1986. Biogéographie évolutive, Masson, Paris

Bourlière, F. 1961. Symposium sur les déplacements saisonniers des animaux. Introduction. Rev. Suisse Zool. 68, 139–143

Brosset, A. 1982. Le peuplement de Cyprinodontes du bassin de l'Ivindo. Rev. Ecol. (Terre Vie) 36, 233–292

Connell, J.H. 1978. Diversity in tropical rainforests and coral reefs. Science 199, 1302–1310

Davies, P.S. 1977. Carbon budgets and vertical zonation of Atlantic coral reefs. Proc. 3rd Int. Coral Reef Symp. 1, 391–396

Duellman, W.E., and Trueb, L. 1986. Biology of Amphibia, McGraw-Hill, New York

Eckert, G.J. 1985. Settlement of coral reef fishes to different natural substrata and at different depths. Proc. 5th Int. Coral Reef Symp. 5, 385–390

Emery, A.R. 1973. Comparative ecology and functional osteology of fourteen species of damselfish (Pisces: Pomacentridae) at alligator reef, Florida Keys. Bull. Mar. Sci. 23, 649–770

Gentry, A.H. 1982. Patterns of neotropical plant species diversity. Evol. Biol. 10, 1–84

Gentry, A.H., and Dodson, C. 1987. Contribution of nontrees to species richness of a tropical rain forest. Biotropica 19, 149–156

Gill, D.E. 1986. Individual plants as genetic mosaics: ecological organism versus evolutionary individuals. In: Crawley, M.J. (ed.), Plant Ecology. Blackwell, Oxford, pp. 321–343

Goulding, M. 1980. The Fishes and the Forest. Explorations in Amazonian Natural History, Univ. of California Press, Berkeley

Grassle, J.F. 1973. Variety in coral reef communities. In: Jones, O.A., and Endean, R. (eds.), Biology and Geology of Coral Reefs (2. Biology 1.), Academic Press, London, pp. 247–270

Harper, J.L. 1977. Population Biology of Plants, Academic Press, London

Hatcher, B.G. 1988. Coral reef primary productivity: A beggar's banquet. Trends Ecol. Evol. 3, 106–111

Holloway, J.D. 1987. Macrolepidoptera diversity in the Indo-Australian tropics: geographic, biotopic and taxonomic variations. Biol. J. Linn Soc. 30, 325–341

Larkum, A.W.D. 1983. The primary productivity of plant communities on coral reefs. In: Barnes, D.J. (ed.), Perspectives on Coral Reefs, Australian Institute of Marine Sciences, Townsville, pp. 221–230

Lavelle, P. 1986. Associations mutualistes avec la microflore du sol et richesse spécifique sous les tropiques: l'hypothèse du premier maillon. C.R. Acad. Sci. Paris 302, Série III, 11–13

Leighton, M., and Leighton, D.R. 1983. Vertebrate responses to fruiting seasonality within a Bornean rain forest. In: Sutton, S.L., Whitmore, T.C., and Chadwick, A.C. (eds.), Tropical Rain Forest: Ecology and Management, Blackwell, Oxford, UK, pp. 181–196

Lewis, J.B. 1977. Processes of organic production on coral reefs. Biol. Rev. 52, 305–347

Lindgren, D. 1972a. The temperature influence on the spontaneous mutation rate. I. Literature review. Hereditas 70, 165–178

Lindgren, D. 1972b. The temperature influence on the spontaneous mutation rate. II. Investigation by the aid of waxy mutants. Hereditas 70, 179–184

Lowe-McConnell, R.H. 1975. Fish Communities in Tropical Freshwaters: Their Distribution, Ecology and Evolution, Longman, London

Lowe-McConnell, R.H., 1987. Ecological Studies in Tropical Fish Communities, Cambridge Univ. Press, Cambridge

May, R.M. 1986. Biological diversity. How many species are there? Nature 324, 514–515

Morel, G., and Bourliére, F. 1962. Relations écologiques des avifaunes sédentaire et migratrice dans une savanne sahélienne du Bas Sénégal. Terre Vie 16, 371–393

Nikolsky, G.V. 1963. The Ecology of Fishes, Academic Press, London

Ogden, J.C. 1985. Herbivore-plant interactions on coral reefs. Conclusion. Proc. 5th Int. Coral Reef Symp. 4, 79–80

Price, P.W. 1984. Alternative paradigms in community ecology. In: Price, P.W., Slobodnikoff, C.N., and Gaud, W.S. (eds.), A New Ecology. Novel Approaches to Interactive Systems. Wiley, New York, pp. 353–383

Richard, G. 1985. Fauna and Flora. A first compendium of French Polynesian sea-dwellers. Proc. 5th Int. Coral Reef Symp. 1, 379–518

Roosmalen, M.G.M. van. 1985. Fruits of the Guyanan Flora, Institute of Systematic Botany, Utrecht Univ.

Sale, P.F. 1984. The structure of communities of fish on coral reefs and the merit of a hypothesis-testing, manipulative approach to ecology. In: Strong, D.R. Jr., Simberloff, D., Abele, L.G., and Thistle, A.B. (eds.), Ecological Communities: Conceptual Issues and the Evidence, Princeton Univ. Press, Princeton, NJ, pp. 478–490

Sale, P.F. 1985. Patterns of recruitment in coral reef fishes. Proc. 5th Int. Coral Reef Symp. 5, 391–396

Serra, J.A. 1968. Modern Genetics, Vol. 3, Academic Press, London

Silberbauer-Gottsberger, I., and Eiten, G. 1983. Fitosociologia de um hectare de cerrado. Brazil Forestal 54, 55–70

Terborgh, J. 1985. Habitat selection in Amazonian birds. In: Cody, M.L. (ed.), Habitat Selection in Birds, Academic Press, London, pp. 311–338

Terborgh, J. 1986. Community aspects of frugivory in tropical forests. In: Estrada, A., and Fleming, T.H. (eds.), Frugivores and Seed Dispersal, Dordrecht, Junk, pp. 371–384

Tilman, D. 1982. Resource Competition and Community Structure, Princeton Univ. Press, Princeton, NJ

Toft, C.A. 1985. Resource partitioning in amphibians and reptiles, Copeia, 1985, 1–21

Vitousek, P.M., and Sanford, R.L. Jr. 1986. Nutrient cycling in moist tropical forest. Annu. Rev. Ecol. System. 17, 137–167

Wanders, J.B.W. 1976. The role of benthic algae in the shallow reef of Curaçao (Netherlands Antilles). I. Primary productivity in the coral reef. Aquatic Bot. 2, 235–270

Whittaker, R.H. 1977. Evolution of species diversity in land communities. Evol. Biol. 10, 1–87

Wilson, E.O. 1987. The arboreal ant fauna of Peruvian Amazon forest a first assessment. Biotropica 19, 245–251

Subject Index

Taxonomic Index